INTO
THE
STORM

INTO
THE
STORM

TWO SHIPS,
A DEADLY HURRICANE,
AND AN EPIC BATTLE
FOR SURVIVAL

TRISTRAM
KORTEN

BALLANTINE BOOKS   NEW YORK

To my mother, Patricia, who taught me to love words; my father, Chauncey, who taught me to love the ocean; and Rosario, Kiara, and Niamh, who taught me to love deeply.

CONTENTS

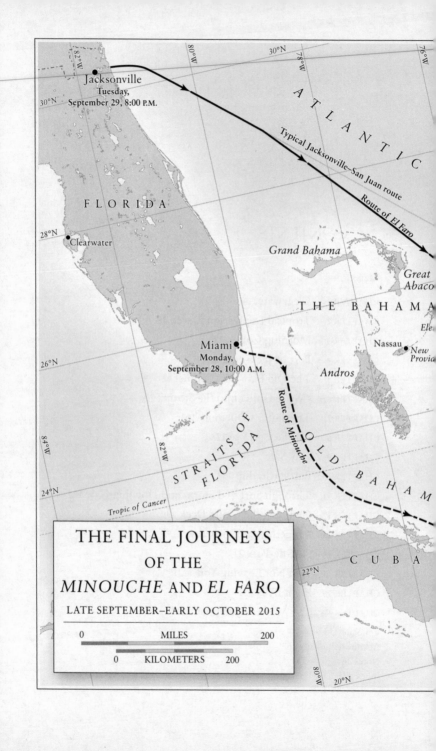

THE FINAL JOURNEYS
OF THE
MINOUCHE AND *EL FARO*

LATE SEPTEMBER–EARLY OCTOBER 2015

| 0 | MILES | 200 |

| 0 | KILOMETERS | 200 |

Jacksonville
Tuesday,
September 29, 8:00 P.M.

Clearwater

FLORIDA

Miami
Monday,
September 28, 10:00 A.M.

Grand Bahama

Great
Abaco

THE BAHAMA

Ele

Nassau • New
Provia

Andros

A T L A N T I C

Typical Jacksonville–San Juan route

Route of El Faro

Route of Minouche

STRAITS OF
FLORIDA

O L D   B A H A M

Tropic of Cancer

CUBA

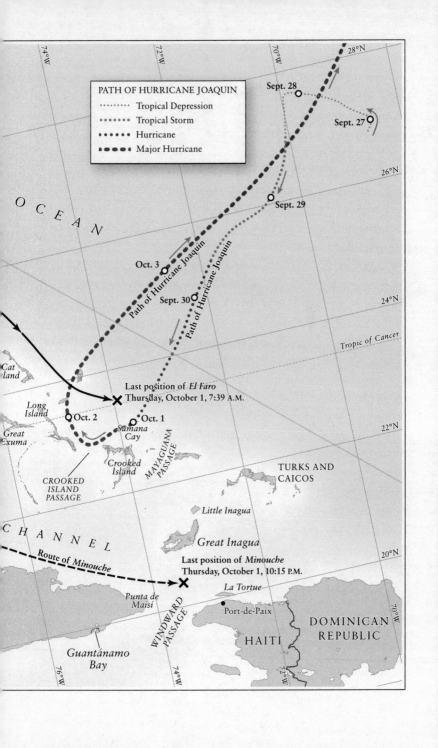

PATH OF HURRICANE JOAQUIN

········· Tropical Depression
•••••••• Tropical Storm
•••••••• Hurricane
▬▬▬▬ Major Hurricane

28°N

Sept. 28

Sept. 27

26°N

O C E A N

Sept. 29

Oct. 3

Path of Hurricane Joaquin

24°N

Sept. 30

Path of Hurricane Joaquin

Tropic of Cancer

Last position of *El Faro*
Thursday, October 1, 7:39 A.M.

Cat
land

Long
Island

Oct. 2

Oct. 1

Great
Exuma

*Samana
Cay*

22°N

*Crooked
Island*

MAYAGUANA PASSAGE

TURKS AND
CAICOS

CROOKED
ISLAND
PASSAGE

Little Inagua

C H A N N E L

Great Inagua

Route of *Minouche*

Last position of *Minouche*
Thursday, October 1, 10:15 P.M.

20°N

La Tortue

*Punta de
Maisi*

WINDWARD PASSAGE

Port-de-Paix

DOMINICAN
REPUBLIC

HAITI

*Guantánamo
Bay*

76°W   74°W   72°W   70°W

INTO
THE
STORM

Chapter 1
CLEARWATER

The C-130 cargo plane sat on the tarmac, a hulking, high-winged, metal-sheathed beast of burden, painted white and orange, the belly so low to the ground it obscured its own wheels. The predawn sky above was dark, but not still. It never is. Currents of air continued journeys that had started continents away, where they twisted and flowed across mountains and deserts, then forests and oceans, pulled and pushed by troughs and ridges of low and high pressure, deflected and guided across an invisible topography, propelled by the sun's heat and the planet's motion. Cool air descended while warm air rose. Within the currents, celestial gases drifted through a transparent skyscape of peaks and valleys—nitrogen, oxygen, carbon dioxide, ozone, and water vapor—redistributed according to their weight up to the stratosphere. The moisture formed into clouds—cirrus, altocumulus, cumulus, stratocumulus, the towering and formidable cumulonimbus—that dissolve and re-form. Below this invisible ballet, in the emergent morning of September 17, 2015, the plane waited on a runway at the U.S. Coast Guard's Air Station Clearwater, on Florida's west coast. Nearly twenty thousand horsepower spread over four turboprop engines, waiting to ride up and surf those clouds.

Across town, Ben Cournia woke in the blue-black of early morning to a quietly chirping alarm. He hit the button quickly and eased his long body up, careful not to disturb his wife, Lindsay, as she slept next to him. She'd have to get up in a couple of hours and go through the tumult of getting the kids to school by herself—making their lunches, packing their backpacks, and hustling them along through teeth brushing and getting dressed. Lord knows, she needed her rest.

Cournia quietly made his way to the bathroom and bent his tall frame over the sink as he turned on the water, his skin pale in the bathroom light, his long arms ropey with muscles and veins. He had thick brown hair, cut short per regulations, and his eyes were set back under a prominent brow. Cournia's swimsuit hung from the shower curtain rod, still damp from the swim in the pool last night. This is their tradition. The night before Daddy leaves for his two-week deployments, the kids get an after-dinner swim in the pool.

As Lindsay cleared the table, Lucy, six, and Lincoln, three, had run squealing to their rooms and wiggled into their suits. Ben also changed into his, then went out and knelt by the pool's edge, furrowing his brow, pretending to fix something. The kids came running out, the soft pads of uncalloused little feet thup-thupping over to him, and—splash!—they pushed Daddy into the pool. *Oh no, he's drowning! Daddy needs help!* The kids jumped in for the rescue. They glided under the water like chubby little dolphins. His kids can swim, Ben has made sure of that. Of course, they rescued Daddy. Afterward, they showered and got ready for bed. Ben tucked them in.

I'll be gone for a little bit, he told them each.

How long, Daddy? Lucy asked.

About two weeks.

Are you going to save people?

Yes.

Okay.

Lucy and Lincoln were learning the rules and rhythms of a Coast Guard family: A parent sometimes has to go away to help others. Their father is a U.S. Coast Guard aviation survival technician, otherwise known as a rescue swimmer. His job is to save people in the worst conditions possible. There aren't many reasons good and clear enough for a child to accept a parent's absence, but rescuing people in danger is one of them.

In the dark of the morning, Cournia packed his toothbrush. He changed into his one-piece green flight suit, freshly washed the night before, and headed down to the kitchen. He brewed a cup of coffee, which he needed to shake the last vapors of sleep from his brain before he started packing the cooler. He didn't want to forget anything, *anything.* He slid the cooler over to the fridge and opened the freezer door. First to go in were the packages of frozen vegetables: green beans, scalloped potatoes, and broccoli, as well as cookie dough to make chocolate chip cookies in his hooch on his off days. Then he opened the refrigerator door to load perishables: chicken breasts, sirloin steaks, pork tenderloins, sliced cheeses and deli meats for sandwiches. Lots of fruit: oranges, bananas, berries. Cournia makes his living off his body, and he is acutely aware of what fuel is needed for it to work at peak efficiency. The calories in his diet should break down as follows: 40 percent carbohydrates, 40 percent proteins, 20 percent fats. To achieve that goal, every meal during his two-week stint is planned out with military precision. He closed the cooler, latched it, and carried it out to the car.

This house, four bedrooms with the backyard pool on a quiet cul-de-sac, is the couple's first home. Cournia's last posting was on a snow-covered island in Alaska, where the family lived on base. But there's no base housing in Clearwater, so they scrimped and saved, and now they have their piece of normalcy, their piece of peace. Cournia loaded the cooler and his duffel bag into his Ford Taurus, slid into the seat, and started the engine. He steered south toward Air Station Clearwater, the house in the rearview mirror slipping into the darkness behind him.

Versions of this ritual were playing out across Clearwater and neighboring Tampa. At flight mechanic Joshua Andrews's home, his wife, Marleen, got up to make coffee while Josh, unflappably friendly with prominent ears and an easy smile, started packing his cooler. Josh is from Texas and takes his grilling seriously. He had marinated his meats the day before, sealed them in plastic bags, and put them in the freezer. He had also bought enough food for an extra week or two. Marleen typically questioned this. *Honey, are you sure you need all that?* He always gave the same reply: *Prepare for the worst, hope for the best.* If something went wrong, a plane broke down or a big storm came, not unheard of this time of year, and they were stuck on deployment longer than usual, he wanted to be ready.

They, too, had let the kids, Ashton, nine, and Leah, five, stay up later than usual to spend extra time with Dad. They'd all piled onto the couch and watched *The Flash,* the superhero show on the CW. When Josh put Leah to bed, she asked why he had to go. *In case somebody needs my help,* he told her. Unlike the Cournia kids, she never accepted this answer and didn't this time, either, judging by the frown on her face, which meant she would probably spend most of the nights while he was gone in bed with Mom.

The kids insisted on saying goodbye in the morning, despite the early hour, so after Josh had packed his bags and dragged them to the door, Marleen went to rouse Ashton and Leah. Even the three dogs got in on the act, swarming the doorway. Cody, the oldest, positioned himself by the threshold, ready to make a mad dash for the truck. Then the kids, groggy, emerged. Ashton hugged Dad, who reminded his son to behave, listen to his mother, and watch out for his little sister. He bent down and scooped up Leah, still warm from bed, nuzzled her round cheek, and reluctantly put her down. She snuck behind him and slipped a picture of a heart she had drawn into one of his bags. "I'll miss you, Daddy," she'd written.

The Post family had a unique situation: Both Rick and Rachel were Coast Guard helicopter pilots. They had a daughter who was about to turn one, so when one of them was deployed, the other was left behind with their infant and a full-time job. Support from friends and their church helped, people who could look after the baby, but it was tough. Rick and Rachel liked to joke that the real tour of duty was staying home. At least on deployment they could get a solid night's sleep.

This morning it was Rick's turn to ship out. His first alarm went off before 5:00 A.M. The couple were such heavy sleepers they used three alarms. The goal was to get up before the big alarm, the third and final one with the metal bells, started clanging in the bathroom, forcing someone to get out of bed and shut it off. Rick made it just in time. He was a thin twenty-eight-year-old, with an angular face, narrow eyes, and a sharp nose that gave him an avian appearance, which was appropriate for a pilot. Rick looked like he never stopped concentrating.

After he had thrown his gear bag and food into their big Suburban he went back in to change into his flight suit. Rachel was just starting to wake up. He kissed her goodbye, then went to get one last look at the baby, grateful he'd be back in time for her first birthday.

At the air station the guardsmen parked in the long-term lot adjacent to the tarmac, where the C-130, officially the HC-130H Hercules, was being readied for flight. The plane was gleaming white, with the Coast Guard's signature safety-orange sash painted right behind the cockpit, overlaid by the service's coat of arms, a stars-and-stripes shield on top of two crossed anchors. The tail was also orange, with a single star emblazoned on it. Before each flight, the hulking aircraft gets not just a full working over to assess all systems but a thorough washing and cleaning. Any mud on the flaps, any chipped paint on the wings, would be an unacceptable symbol of shoddiness.

The aging workhorse of the Coast Guard's air fleet, the Herc is 99 feet long and 38 feet high with a 132-foot wingspan. It can fly 2,500 miles before needing to refuel. The plane is not fast, and it's not sleek, but it can land on a rutted field, launch with only 800 feet of runway, and haul anything anywhere. Hercs have been used to drop bombs, transport troops, and catch satellites. In sixty years of service it's been a flying ambulance, midflight refueler, hurricane hunter, fire tanker, and search and rescue platform. It is tough and versatile, which is exactly how the Coast Guard sees itself.

By far the smallest of the country's military services, the Coast Guard has about 40,000 members, not including reserves. The next smallest is the Marines, with about 190,000. The Navy has 315,000, followed by the Air Force with 330,000, and that great lumbering behemoth, the Army, with 487,000.

Yet the Coast Guard is busy whether we are at war or not. Not only does it rescue people in distress at sea as well as during floods and other natural disasters, it is also tasked with securing all ports; enforcing maritime laws; patrolling our coasts and waterways; protecting underwater marine resources; patrolling against illegal fishing; inspecting commercial and recreational vessels; and a host of other law enforcement jobs. Because its numbers are so small, the service can't afford to have its personnel specialized too narrowly. In the Coast Guard, everyone juggles several duties. On an air station such as Clearwater, the base commander, communications officers, and office workers are all expected to fly patrols.

The Hercules requires a crew of seven and has a cavernous interior, with rows of parallel roller tracks running the length of the fuselage floor. This design allows it to be quickly converted from a cargo plane to a troop transport. Special pallets that lock into the rollers allow up to fifty-one thousand pounds of cargo, such as supplies and vehicles, to be loaded onto the rear ramp and slid inside. (The width of the fuselage was determined by drawing a circle around an M551 Sheridan tank.) On rescue missions, life rafts and dewatering pumps can be rolled into the plane and dropped on or near struggling ships. During the rush to contain a tanker gushing oil into the sea, the Herc can carry dispersants to help control the spill. Then, just as easily, it can be transformed into a passenger plane by swapping out the cargo pallets for pallets with seats bolted onto them. The seats get rolled up into the belly of the plane just like the life rafts would, and in no time you've got seating for ninety-two.

On this day, the Herc's duties were split between passengers and cargo. In the hour or so before they were scheduled to take off, Ben Cournia and the others lugged their gear bags and coolers to the rear of the plane and tossed them onto a pallet. Then they headed to the mess hall to get some breakfast. After he ate, as he walked back to the plane, Cournia pulled out his cellphone and texted Lindsay. She'd be up by now.

Stuff is loaded, he wrote. *Getting ready to leave. See you in a couple of weeks.*

OK, she shot back. *Be safe.*

No problem there, he thought to himself. In the five years he'd been going on these deployments, he'd never once had to worry about being safe.

By now the mid-Florida sun was just beginning to spread orange tendrils of light across the airfields. Birds trilled in the palm trees as the Coasties gathered outside the plane. Flight crews in their green one-piece suits and ground crews in their blue ODUs, or operational dress uniforms—blue tunic, blue cargo pants, and blue visored caps—hoisted their gear onto the pallets and assembled in a line to climb the steps on the plane's left side, just behind the cockpit. The morning was still cool, and the sharp smell of jet fuel filled the air. Joining Cournia, Post, and Andrews as they boarded the Herc was Lieutenant Dave McCarthy, a helicopter pilot. McCarthy, the senior officer for the next two weeks, was the final member of their four-man Jayhawk helicopter crew, one of two such crews that would rotate duty on this deployment.

The four men made up a pretty good snapshot of the Coast Guard in terms of demographics (the service is overwhelmingly white) and disposition. There was no hoo-ahing, no swagger. They were soft-spoken men, who, when they did speak, were impeccably polite and used precise language. They didn't get into drunken brawls while off duty. They went home to their families.

It's hard to say whether they joined the Coast Guard because of their steady demeanor, or if the Coast Guard developed this in them. Their motivations for joining varied. Post had always wanted to be a pilot and entered the Coast Guard Academy straight out of high school. Andrews was searching for a way out of a series of jobs that were going nowhere. Cournia was looking for adventure and a challenge, and McCarthy was hit with an epiphany while visiting the Vietnam Veterans Memorial in D.C. They had all been in ten years or more. They were serious professionals who spent a lot of time training to remain calm in dangerous situations. They believed in what they were doing, and they were good at it.

The Herc may have gleamed like a new toy on the outside, but inside it was a testament to raw function. Tubes, valves, and wires snaked over a patchwork of insulated quilting that covered the cabin. If you took your average passenger plane and tore out the inside walls, along with the carpet and anything else that made it comfortable, this is what would be left. There was a dispenser full of yellow spongy earplugs by the door because the interior is not soundproofed against the roar of the engines. And God forbid you have to go to the bathroom. The only way to do that is to climb over the cargo and pull a folding toilet seat down from the wall, which the ground crew loading the plane had probably jammed shut by piling gear against it. If you managed to pull the seat down and desired even a shred of privacy, you'd have to grab the corner of a big canvas curtain and hold it up with one hand while attending to business with the other—hoping the plane didn't hit any air pockets while you were busy.

Cournia settled into his seat. All around him guys were slipping earbuds in, cracking books, opening computers to watch movies, or just nodding off and settling in for the three-hour flight to what was surely one of the loneliest Coast Guard outposts in the service: a glorified sandbar in a remote part of the Bahamas called Great Inagua Island. That's a grandiose name for what is essentially a sunbaked, mosquito-infested outcropping of sand-covered limestone overrun by feral donkeys. But it's strategically located, and so the base is manned at all times by two helicopter crews and the ground personnel that keep them aloft.

The Bahamas assignment was a bit of a change for Cournia, who'd grown up in northern Minnesota and spent four years stationed in Kodiak, Alaska. There he'd routinely gone on patrol and search and rescue missions in subzero weather, often in the dark. By comparison, the Great Inagua posting was a little quiet for his taste. In five years of deployments to the placid subtropics, he'd had only two search and rescue cases, and one of them was to simply medevac someone off a recreational boat. A steely Midwesterner who'd made it through rescue swimmer school (less than half of the candidates graduate), Cournia had earned his spot as one of the Coast Guard's most elite operatives. His skills and conditioning were on par with a professional athlete. Yet he was thirty-six, getting up there for a rescue swimmer, and he worried he'd never get another chance to make his mark.

On Great Inagua, the pilots got to fly their beloved MH-60 Jayhawk helicopters. The flight mechanics got to tinker with whatever it was they tinkered with. A rescue swimmer like Cournia? He'd be lucky if he got his feet wet. He'd probably spend most of the deployment riding in the back of the chopper, working the radio and the lights during patrols for smugglers. He'd have to work on his rescue swimming skills by himself, hitting the weights and getting some running and swimming in. At the end of the day, at least he could reward himself by grilling steaks and baking cookies.

A few seats away, McCarthy was much more amped about his next two weeks. He had arrived at Clearwater only about six months ago, and this was his first deployment to Great Inagua. He was going to be the senior aviator, which meant mentoring the less experienced pilots. He looked forward to some new challenges.

The unknown was nothing new for McCarthy. Though his life by the age of thirty-six was stable and conventional—wife, four-year-old daughter, Coast Guard career—his upbringing had been anything but. His father had been a stage magician and entrepreneur, but financial difficulties had dogged the family and they'd moved repeatedly, from Arizona to New Jersey to Florida, where his father opened magic stores. When McCarthy was young, the family lived for a stint in a converted school bus in Orlando. McCarthy missed half of fifth grade during the upheaval. He distilled this experience down to a single phrase: You make the best of what you have. He was a man who was intimately familiar with the world's potential for chaos, and he did everything within his power to keep it at bay. That was why the life of a pilot agreed with him. You completely controlled a machine that was hovering in the most unpredictable environment of all: the air. There was magic in that, too.

As the Herc prepared for takeoff, McCarthy reached down and adjusted the bulge in his flight suit's left ankle pocket, where he had crammed a stuffed turtle that belonged to his daughter. She had been so upset at his leaving that he offered to take the turtle with him and send back pictures of the adventures they had together. She thought this was a great idea. Now he pressed play on a country music mix—Kenny Chesney, Toby Keith—and leaned his head back to enjoy the ride.

The plane hummed, then roared as the four 4,900-horsepower engines started to spin the propellers. Inside, men stuffed in ear-plugs to drown out the deep thrumming. As air rushed under the wings and the plane lifted off the runway and sped over Old Tampa Bay, the guardsmen could look down at Air Station Clearwater spread out below them: seventy-two acres of runways, a cluster of barracks-style buildings that look like they haven't changed much since the base was commissioned in 1934, and four large hangars roomy enough to house the C-130s and Jay-hawks. It was the service's busiest air station, covering a region that stretched from the Gulf of Mexico north into the Atlantic and southeast into the Caribbean. Sewn onto all the flight suits was the station's motto: "Anytime, Anywhere." It was part brag, part promise.

The Hercules flew southbound at twenty thousand feet, then east across Florida. The blue of Lake Okeechobee and the soft greens of the Everglades soon gave way to the paved-over sprawl of the West Palm Beach–Fort Lauderdale–Miami megalopolis, then the lazurite blue of the Florida Straits to the Great Bahama Bank. Beneath them lay one of the busiest waterways, not just in the country, but in the entire Western Hemisphere—a vast ma-rine superhighway clogged with vessels of every shape and size carrying all manner of consumer goods: container ships piled with metal boxes; bulk cargo ships hauling raw materials such as molasses, rice, and lumber; general cargo ships, their holds packed with cars and trucks and other freight; oil tankers; massive cruise ships (Miami is the busiest cruise-ship port in the world); grease- and rust-streaked fishing trawlers. In a more sinister vein, there are the go-fast motorboats that offload drugs from mother ships and scurry them ashore, along with the various craft used by human smugglers bringing job- and freedom-seeking Cubans, Dominicans, and Haitians into the United States. All of these vessels ferry their goods back and forth in the continued expectation of smooth, profitable journeys. Add to this commercial traffic the extraordinary density of pleasure boats, both sail and motor, that speckle the seas off the southeastern United States and into the Caribbean: The Coast Guard is expected to watch over all of this.

Among the vessels plying their trade in these shipping lanes were two cargo ships that sailed routine runs between the Caribbean islands and the Florida ports that supplied them with the groceries and sundries of daily life. The *Minouche*, 230 feet long, shipped out of a small dock along the Miami River to Port-de-Paix, Haiti. *El Faro*, 790 feet, sailed out of the Port of Jacksonville, at the other end of the state. It's possible the two ships had passed each other along the Old Bahama Channel, the well-traveled shipping lane that runs between Florida and the Bahamas, during the few times that *El Faro* deviated from its more northerly route to avoid bad weather. The vast majority of these journeys were uneventful, especially in the summer months when the ocean this close to the equator often lies so flat it's referred to as Lake Atlantic.

Both ships were veterans. The *Minouche* had been built thirty-five years earlier, *El Faro* forty. Both had seasoned captains who had spent their whole lives on the water and had sailed much more challenging seas all over the globe. Despite these similarities, the two ships were from utterly different economic worlds. *El Faro* loaded at a modern port where mechanized gantry cranes lifted uniformly sized containers onto the deck, filling the entire ship in a day using computer programs that plotted the weight distribution. The scrappier *Minouche* still loaded loose cargo—

boxes, crates, and industrial-weight plastic bags—by hand with the help of a boom, or derrick, on its deck. The derrick lifted the cargo in a big net, and stevedores carefully distributed the load in the ship's holds and across the main deck, a process that took days.

But the weather and water make no distinction between the rich and poor, and the best engineering and electronics matter little against waves that can break with thousands of pounds of pressure per square foot, or winds that can blow so hard they'll lift a man off the deck. The ocean is, and always has been, the great equalizer.

The C-130 banked hard over the landing strip on Great Inagua to scare off the wild donkeys that sometimes stray onto the tarmac, then came in for the landing. The sky was blue, the sun bright yellow, almost white, and the heat enveloped the men like a warm bath as soon as they stepped off the plane. The island's famously aggressive mosquitoes greeted this fresh meat with enthusiasm. Everyone hustled to retrieve their gear from the rear of the plane and hurried to their housing units, called hooches, before their blood was drained. McCarthy realized he'd forgotten his flight gear—helmet and gloves. He chanted another favorite mantra—"Adapt and overcome"—and jogged after one of the pilots returning to Clearwater to bum a set. Then he followed everyone else to their hooches.

As the men headed in to unpack, they passed a hangar on their right, a fortress-like building that housed the two Jayhawks and the small tractors, called mules, that towed the helicopters onto the runway. Inside was a communications room, a storage area, and a gym. A small bronze plaque with relief letters was mounted on the outside wall:

DESTROYED BY HURRICANE IKE ON SEPTEMBER 7, 2008
REBUILT AND DEDICATED ON MARCH 21, 2013

Ike was a brute that had formed from a low-pressure wave off the west coast of Africa on August 28, 2008. The storm began as a roiling mass of thunderheads south of the Cape Verde Islands, off Senegal, before developing into a Category 3 hurricane that barreled into the islands of Turks and Caicos. Then it headed straight for Great Inagua at 110 knots, or 126 miles per hour. And it wasn't alone. Ike was one of three back-to-back hurricanes, including Gustav and Hanna, to spin through the region within a hellish ten-day span.

The plaque was a small reminder not to be lulled by the sunny subtropical weather.

The base on Great Inagua is part of an international effort called Operation Bahamas and Turks and Caicos, reduced in militarese to OPBAT, which is primarily aimed at stopping drug smugglers. OPBAT was started in 1982, after U.S. officials discovered that Colombian drug trafficker Carlos Lehder had essentially taken over an entire island in the Bahamas called Norman's Cay to use as a base for smuggling cocaine into the United States. His operation came complete with an airstrip and armed guards patrolling the beaches. After bringing Lehder down, the feds crafted an agreement with the Bahamas and neighboring Turks and Caicos to set up two bases, one on Andros Island and one on Great Inagua. DEA agents work alongside the Coast Guard and Royal Bahamas Defence Force (RBDF) agents to jointly patrol the surrounding waters for smugglers. As one DEA agent described it in a 2016 *Coast Guard News* article, "The DEA is the primary intel, the Coast Guard has the assets and the country agents are the jurisdiction. You can't do the operation with just one."

Great Inagua was chosen for a reason. It is the most southerly island of the Lucayan Archipelago, the other name for the Bahamas. It sits just north of the Windward Passage, the strait between Cuba and Hispaniola, the island comprising Haiti and the Dominican Republic. The Windward Passage is one of the most direct water routes to the United States from South America. Having a base to watch the ships that sail through it is invaluable.

The island itself is obscure and sparsely populated, but not small. It stretches fifty-five miles from its western tip to its eastern tail and covers nearly six hundred square miles in total. The interior is cored by a huge brackish lake. Great Inagua's nine hundred full-time human residents are far outnumbered by the wildlife, which include those feral donkeys, left over from French colonists hundreds of years ago, and a flock (or, in the extraordinary parlance of the birding world, a "flamboyance") of some eighty thousand pink West Indian flamingos.

The birds thrive there, in part, because of the island's large salt-drying reservoirs, owned by the Morton Salt company, which produces about a million pounds of sea salt a year on Great Inagua. As the sun and wind evaporate the water in the reservoirs, leaving the salt behind, an algal mat develops on the surface. Brine shrimp gather to eat the algae, helping to clean the water. This attracts the flamingos, which gather to gorge on the shrimp, giving the birds their pink color. As a result, the flamboyance on Great Inagua has grown to be the largest in the Caribbean.

The drug patrols, too, are often fruitful. But the Coast Guard gets its identity from saving people, not arresting them. Flying over blue water, peering through the sites of a machine gun looking for smugglers, is not necessarily how some Coasties would prefer to define their jobs. They tend not to complain about it too much—an order is an order—but this attitude sometimes peeks through the official veil. The service's publication, *Coast Guard Outlook*, hinted at this in its 2015–16 edition: "The guns that were mounted later on helicopters were meant to disable the engines of smugglers' 'go-fast boats'—not to shoot people. The 'war on drugs' was, and is, controversial and at least some Coast Guard members say they derive more satisfaction from rescue duty."

———

After dropping off the fresh deployment, the big C-130 Hercules took off with the crews who had just finished their two-week stints and were headed back to Clearwater. But first it would deliver a shift change to the single-helicopter base on Andros Island.

Meanwhile, the Great Inagua crew settled into their concrete-block hooches. The quarters resembled dorm rooms at a state university. Blocky wood chairs and coffee tables in the living room, single beds under single windows in each of the individual bedrooms, four per hooch. During their shifts the Coasties would be kept busy flying patrols in the helicopters: a pilot, co-pilot, flight mechanic, and rescue swimmer in each four-man crew, along with a DEA agent and maybe a RBDF agent along for the ride. They'd conduct searches based on intelligence from ships at sea or from DEA sources and communicate with Coast Guard cutters, also on routine patrol in the waters. Looking down, they'd scan what seems like an endless blue, hovering over ships around the Windward Passage, looking for anything suspicious.

During their off hours the guardsmen were free to take the base's motorboat out for some recreational fishing. The men had recently been getting into spearfishing in the glass-clear Bahamian waters. No spear guns allowed. In the Bahamas, diving underwater for your catch requires you to use a more primitive pole spear or Hawaiian sling, giving the fish a sporting chance. Even with that challenge, the men could eat well. Mangrove and red snapper, grouper, hogfish—they'd all end up on the base's grills before the deployment was up. And at this time of year, the tail end of summer, the very beginning of fall, the water typically lay flat, calmed by the subtropical sun and winds that were almost *too* gentle.

Or, as Cournia put it, "It can be pretty boring."

Chapter 2

TROPICAL DEPRESSION ELEVEN

About ten miles up the Miami River from where it empties into Biscayne Bay, past the shimmering office towers of downtown Miami and the low-rise streets of Little Havana, northwest past the modest, squat stucco homes of Allapattah, right where the railroad tracks run by a messy sprawl of scrap metal yards, Captain Renelo Gelera woke up in his cabin on the general cargo carrier MV *Minouche* ("MV" for "motor vessel"). The name was French slang for "kitty cat." Gelera washed and combed through his thick black hair, then said his daily prayer, etched into a metal bracelet on his wrist: *Padre Nuestro, que estas en el cielo, santificado sea tu nombre. . . .* He prayed that his ship and crew be kept safe from the wind and waves. He prayed for his wife and children in Guatemala. He prayed for his mother and siblings in the Philippines. And he prayed that he and his crew would set sail soon, because he was going stir-crazy waiting around in port.

He climbed down the stairs from his captain's quarters, a small bedroom attached to an office, and stepped out on deck. Gelera is short, trim, and nimble, physically well suited to life in the narrow hallways and passages on a ship. He looked up at the dark purple of the pre-sunrise sky. A light breeze cooled the air. That whisper of wind would be gone by noon, Gelera expected, replaced by a searing sun and a thick layer of humidity. It was Sunday, September 27, 2015, which in the subtropics was still as broiling hot as any day at the height of summer. He looked down at the wharf, piled with the last boxes, crates, and cars waiting to be loaded into the ship's holds and onto her deck. His mind began processing all the tasks that lay before him: oversee the cargo lashings, stock up on supplies, check the fuel levels. Then he turned and headed into the galley to have breakfast.

At fifty-eight, Gelera has been a shipmaster a long time, a sailor for even longer. He left home at sixteen to go to maritime school in Cebu City, Philippines, and essentially grew up on the decks of ships. Since then, he has become a nomadic citizen of the world. Maybe more accurately, a citizen of the waves. He is a polyglot, fluent in English and Tagalog and able to hold his own in Haitian Creole, Spanish, Greek, and even a little Mandarin from all the crews he has sailed with. He carries himself with an alert, slightly electric air, smiling and laughing easily, high pitched and loud.

But when it came to his ship, he had the focus of a machine gunner peering through his sights, picking off the tasks the crew needed to attend to in rat-a-tat fashion. Gelera hated wasting time. And a lot of time tends to get wasted when a ship is in port.

The *Minouche* had been docked at Caribbean Ship Services for about two weeks now, waiting for enough cargo to make for a profitable journey. That threshold had finally been met, and the ship's agent had informed Gelera that the ship would sail tomorrow, Monday, September 28, for Port-de-Paix, Haiti. This was the *Minouche*'s regular run, one she made every month.

The *Minouche* was not a large ship at 230 feet, and she was old and streaked with rust. She sat low in the water now, her holds and decks nearly full. The two-boom cargo mast mid-deck would swing the last of the ship's load on board by late afternoon. The fuel truck had stopped by earlier, topping the tanks at five thousand gallons of bunker fuel, the thick and oily residue left over from making gasoline that ships used, enough for the round-trip to Haiti. Gelera could feel the tension lifting. He was a sailor; he needed to be on the water, under way.

Haiti is a beautiful country, with green mountains that reach heights of eight thousand feet and azure water that breaks on beaches of white sand. It is also, of course, poor—the poorest nation in the Western Hemisphere (far worse off than its neighbor, the Dominican Republic) and one of the poorest nations in the world. Whatever can't be grown or made there has to be shipped in. Groceries, bicycles, building supplies—almost all of it comes from somewhere else. But most Haitians can't afford newly manufactured goods, not even from China, so instead there is a brisk trade importing the First World's castoffs: rusted and dented cars, scraps from lumberyards, and mountains of used mattresses that are tossed unceremoniously onto the wharf and left out in the sun and rain until they can be loaded onto the ships. Once in Haiti, their ticking will be stripped off and they will be repacked and sold again.

Miami is home to the largest bloc of the Haitian diaspora, and it's this community that powers this micro-economy, in which hundreds of Haitian and Haitian American merchants use ships like the *Minouche* to send supplies back to *patri*, the motherland. The docks for these ships line the Miami River, nestled in against the rail tracks and metal yards, where East Coast Scrap Metal, Bahia Honda Scrap Metal, and Miami Iron and Metal continue the job of breaking down the vast accumulations of a consumer society. The developed world's scraps loaded down the *Minouche* on this day in late September 2015: tires, windows,

auto parts, vehicles, cooking oil, rice, and eight hundred pairs of
shoes donated by a church in Indiana. But the scene could just as
easily have been a dock in Japan piled high with cargo for a ship
headed to Indonesia or a South Korean vessel bound for the
Philippines. Lightly used products that the citizens of developed
nations discard without a thought—"Oh, time to buy a new
bed/computer/car"—have a whole second life across the sea.

After finishing his breakfast, Gelera walked down the gang-
way to the wharf and out to the parking lot. As usual, he had
waited until the day before departure to buy provisions for the
crew, in order to avoid spoilage. There would be twelve men
aboard the *Minouche* to feed, including himself. He left his chief
officer, a Filipino countryman named Henry Latigo, to oversee
the ongoing loading and lashing of the cargo and headed to the
Toyota Tacoma pickup truck the ship's agent let him use when in
port. Gelera took Jules Julius Cadet, one of his able seamen (a
merchant marine rank for deck workers) and set off on a shop-
ping spree.

Cadet is a large man with sad, rheumy eyes, who comes from
the Haitian island of La Tortue. He has a big family back home,
("maybe eight babies," he said, when asked), and although the
job keeps him away from Haiti for months at a time, his
$1,000-a-month salary is a lifeline in a country where about 60
percent of the population lives on less than $2.40 a day.

Gelera and Cadet drove to the food wholesaler Jetro Cash and
Carry on NW Twelfth Avenue and stocked up in industrial quan-
tities: three 40-pound boxes of chicken wings, two 40-pound
boxes of chicken legs, 80 pounds of turkey drumsticks. They
bought Haitian and Philippine staples, 110 pounds of white rice
and 55 pounds of black beans. (Except for Gelera, Latigo, and
the Dominican engineer, the other nine crew members were
Haitian.) Twenty pounds of hot dogs, ten pounds of pre-sliced
American cheese, and 420 eggs. Plus water, juices, sodas, paper
towels, and plastic forks and spoons. Then they drove to a Winn-

Dixie supermarket, where they picked up about 20 loaves of bread along with ketchup, mustard, vinegar, jam, peanut butter, spaghetti, tomato sauce, and on and on and on, until the pickup's bumper sagged to the curb.

By the time they returned to the dock it was 4:00 P.M. As the crew unloaded the groceries, Gelera inspected the cargo lashings, making sure everything was properly secured. After all, it defeated the whole purpose of shipping if cargo fell overboard mid-journey. Even worse, loose cargo could shift and throw off the boat's steering, costing fuel and time. In rough weather, it could cost a lot more than that.

Hurricane season officially starts on June 1 and runs through November 30, a five-month stretch when the water in the tropical Atlantic heats up sufficiently to fuel the big storms. The National Hurricane Center starts standing watch two weeks before the Atlantic season begins, assigning two-person teams to monitor weather activity around the clock for signs of developing storms in the Atlantic and Pacific, where the season starts May 15. But as man plans, Mother Nature mocks. On May 10, 2015, Tropical Storm Ana blew through South Carolina, making it the earliest tropical or subtropical storm on record to make landfall in the United States.

Scientists are studying this and trying to determine whether the average hurricane season is getting longer. The studies so far have been inconclusive. But in 2016, a meteorologist named Ryan Truchelut analyzed the accumulated cyclone energy of each hurricane season going back to 1979 using a complex mathematical analysis called quantile regression. He determined that although the timeline for energy expended is fairly consistent year after year—"the date[s] at which the Atlantic hurricane season is expected to be 10 percent, 50 percent, or 90 percent complete have not changed significantly since 1979"—early season storms have started blowing about a half day to a day earlier per year. "This means all else being equal, we might expect the first 5 percent of a hurricane season threshold in the Atlantic to be reached about 20 days earlier in 2016 relative to 1979, or approximately July 20 instead of Aug. 10," said Truchelut, who went on to caution against drawing unequivocal conclusions from only thirty-eight years of storm data. In fact, some of that difference may be attributed to improvements in data collection over the decades, since data collected in 2016 is far more detailed than data from 1979. Still, he noted, the likely culprit is global warming, which has heated the water so that the average sea surface temperature is one degree Fahrenheit warmer than it was forty years ago. (Heat transfer from the ocean is the main fuel for tropical storms and hurricanes.)

As Renelo Gelera was completing his pre-trip inspection of the *Minouche,* across town Todd Kimberlain was just starting his shift at NHC headquarters, a squat concrete bunker of a building bristling with satellite dishes and antennae in the suburbs west of Miami. Kimberlain, a hurricane specialist, was working the four o'clock to midnight shift, monitoring the weather over the Atlantic Ocean.

Every time Kimberlain took a seat at his workstation—nine monitors linked to four different computer systems that gave him a God's-eye view over the ocean—he was fulfilling a childhood dream. As a weather-obsessed kid in the 1980s, Kimberlain would listen to his weather radio and call the NHC nearly every day, peppering forecasters with questions about developing weather systems. When he was about eleven years old, he took a tour of the NHC building. Then he took another, and another. Eventually one of the forecasters took Todd under his wing, telling him to stop by on a Saturday and he would show him how to plot analyses. Kimberlain took him up on the offer and visited one Saturday, staying until the director noticed him and kicked him out. Crushed, Kimberlain rode the elevator down in tears. But the forecaster followed and told Kimberlain not to worry.

Let this blow over, he said. *Come back in a few weeks.* So Kimberlain did. This time, they let him stay.

From then on he never considered another career path. During high school he convinced the NHC to give him summer internships. He went on to earn graduate and postgraduate degrees in meteorology and eventually made his way back to the NHC in 2008 to take the spot where he was right now, watching over the Atlantic, looking for signs of trouble.

Kimberlain, who has the look of a perpetual grad student, with a few days' scruff, sideburns, and a mop of thick, brown hair, was scanning the ocean and atmosphere using satellites that beam back images often resembling nothing more than a poor-quality X-ray: light grayish splotches hovering over a darker gray or black background. The previous shift had alerted him to keep an eye on a suspicious thicket of clouds to the northeast, between Bermuda and the Bahamas, that had been spotted a day or two ago.

Kimberlain had no problem finding the cluster of clouds. On his screen, they were now a distinct mass. He zoomed in for a closer look and then created a loop that he could click through frame by frame. That's how he was able to notice the ever-so-slight counterclockwise rotation of the low clouds. The closed circulation of the clouds, coupled with a satellite image of the low-level center, confirmed it in his mind: This was a tropical low-pressure system, which required the NHC to put out an advisory.

The National Hurricane Center is one of four centers that link together to monitor the nation's weather systems. The Weather Prediction Center in Maryland monitors rainfall; the Storm Prediction Center in Oklahoma monitors tornadoes; and the Ocean Prediction Center in Maryland provides marine forecasts, including wind and wave activity. (The NHC's Tropical Analysis and Forecast Branch also provides marine forecasts over the tropics.) All these fall under the National Centers for Environmental Prediction, part of the National Weather Service, which, in turn, is part of National Oceanic and Atmospheric Administration. When weather systems such as a tropical low develop, representatives from all four centers have a conference call before any public advisory is released. The U.S. Navy, which has its own formidable storm-tracking systems, listens in on the call as well to get a heads-up.

The challenge, at this point, was that there was little observational information from passing ships or planes to supplement the satellite images. Kimberlain would have to run this through his boss, James Franklin, the branch chief for the Hurricane Specialist Unit. Kimberlain made his case in a series of text messages to Franklin: the low cloud formation, the cyclonic circulation, and the growing organization of the thunderstorms. After grilling him over it, Franklin approved Kimberlain's recommendation. *If you think it's a system now, go ahead*, he wrote.

Kimberlain was able to determine that northwest winds had blown the thunderstorm activity away from the circulation center, the point around which the clouds were rotating. Meanwhile, a low- to mid-level ridge, an elongated area of relatively high-pressure air north and northeast of the cyclone, was steering the lower half of the depression's circulation northwestward, while increasing winds from the other direction were opposing this motion.

This cloud pattern meant the winds were not likely to get much stronger. Storm systems like uniform winds. Kimberlain was now able to mine the massive amount of computing power at his fingertips—satellite imagery, microwave radar sensors that calculate wind speed and direction by measuring capillary action on the waves—to assemble a more complete picture of what was happening. When he entered the storm's data—location, direction, size—into the computer to be run through several dynamical models, there emerged a consensus that a northerly wind shear, or sudden change in the direction or speed of the wind, would probably blast the low apart by late Monday and scatter it out to sea. The storm had just enough definition, however, that

it needed to be identified and numbered. Kimberlain checked
how many tropical depressions had formed so far that year and
took the next number in the sequence: Tropical Depression
Eleven. Then he wrote his advisory, noting that "global models
show the shear being sufficiently strong enough to result in the
cyclone either becoming a remnant low or dissipating within a
few days."

At 10:00 P.M., Kimberlain organized the conference call to
announce Tropical Depression Eleven. When everyone was on
the line, Kimberlain informed the various agencies that although
there was no real surface data yet—no observations from ships or
planes—the satellites clearly showed that the system was becoming better organized.

But Kimberlain had good reason not to worry about TD
Eleven developing into much. In addition to the high wind shear,
low-pressure systems forming outside the tropical latitudes, like
this one, rarely gain enough strength to become powerful storms.

What the forecasters were really on alert for were lows that
developed from disturbances in the prevailing easterly wind flow
off the coast of Africa, the kind that led to Hurricane Ike back in
2008. This airflow can blow so consistently during the early
summer that fine red dust from Saharan dunes routinely coats
cars and windows in Miami and Fort Lauderdale. Air currents
over Africa blow across the Sahara—a desert the size of the continental United States—and heat up. When this warm, dry air
meets the moister, cooler air over forests and lakes to the west
and south of the African continent, it causes a jet to form. Disturbances in this jet create waves. As these waves of air move
westward over the warm waters of the tropical Atlantic, they
spawn the majority of the Atlantic Basin's hurricanes.

While predicting the behavior of tropical cyclones can never
be an exact science, it's the most accurate it's ever been. Over the
past fifteen years, meteorologists have managed to reduce track
errors for the forty-eight-hour forecast period in the North
Atlantic by 50 percent, largely due to the use of increasingly so-

phisticated satellites and modeling systems and advances in com-
putational weather prediction. Still, forecasters at the National
Hurricane Center will be among the first to tell you their work is
far from flawless.

"Our forecasts are not perfect. We spend a lot of time in our
outreach letting people know how lousy we are," said Franklin.
This is strategic humility. "If people assume our forecasts aren't
susceptible to errors, they can put themselves in danger. Anyone
who lives in a hurricane-prone area understands this. The prob-
ability of success is, by definition, two-thirds of the time."

This is why the forecast track for a hurricane, also known as
the cone of uncertainty, which plots the likely path of the storm
over the next 12 to 120 hours, uses ever-widening circles around
each point on the map as the projection moves further into the
future. The size of those circles is based on the official forecast
errors over the past five years.

Kimberlain handed off Tropical Depression Eleven to the next
shift's forecaster, Dave Roberts, who continued monitoring the
disturbance. In his 5:00 A.M. report, Roberts noted that a "grad-
ual turn toward the northwest is expected Tuesday morning, and
this general motion should continue through Wednesday. . . .
Maximum sustained winds are near 35 mph (55 km/h) with
higher gusts. Some strengthening is possible during the next
couple of days, and the cyclone could become a tropical storm by
Tuesday." In the discussion portion of the forecast, Roberts
noted that the storm would probably dissipate in about three
days because of a wind shear to the north that "should inhibit
any significant strengthening."

Only in retrospect would the forecasters begin to understand
what went wrong. And this is why they stress the error probabilities of any forecast—probabilities that are too easily ignored,
especially by shipping companies and captains with tight delivery
schedules and deadlines to meet.

Chapter 3
MORNING COLORS

The world's oceans and seas are vast and simmering with latent power, the skies above them equally treacherous. The machines men use to stay afloat and aloft, no matter how expensive and technologically advanced, are frail in the face of these unpredictable environments. Captain Richard Lorenzen, Air Station Clearwater's commander, came to work every day with this knowledge at the forefront of his thoughts. He had done three tours of duty in Clearwater, ten years of hurricane seasons. He knew where he, his men, and the machines they used stood in the planet's climatological pecking order.

So when Lorenzen arrived at his office on Monday, September 28, 2015, he was primed to see the world around him as a series of threats. The Coast Guard's motto, after all, is *Semper Paratus*, Latin for "Always Ready." His job was to create a chessboard of possible responses so that whatever may happen, he had positioned the assets at his disposal—people, planes, and helicopters—for a countermove. Lorenzen, a Jayhawk pilot with close-cropped hair the color of steel shavings and a pilot's clear blue eyes, was an operations guy, more comfortable in the cockpit than in his office, and as likely to show up at work in his flight suit as in his uniform pants and shirt. He was coming off a relaxing weekend, so he was curious, maybe even a little excited, to find out what lay in store. He didn't like being out of the loop, a problem that would be quickly resolved at that morning's regularly scheduled Monday briefing.

Just after 7:30 A.M. his executive officer poked his head in the office door. *You ready for the walk, sir?* With that, the two men, joined by the command master chief, left the office building. They marched across the tarmac to the C-130 hangar and made their way down the long, narrow hallway to the law enforcement conference room. As Lorenzen entered, someone barked *Attention on deck!* and everyone stood. "At ease," the captain said as he walked to the head of a boardroom table and sat in his leather swivel chair. On the wall behind him were two display screens.

The meeting began at 7:45 A.M. It used to start at 8:00, but that was also the time for morning colors, the daily ceremony when the flag is raised. Anyone caught outside was expected to stand at attention, face the flag, and salute, which sometimes caused stragglers to show up late to the morning briefing. There was a suspicion on base that the chief petty officers in charge of morning colors might hit the "first call" recording a little early if they saw an officer hustling across the tarmac, just to mess with them.

Lorenzen liked a full room, so he made sure to invite junior officers, pilots, even ranking enlisted members to the briefings. The captain was edging toward retirement the following year and he tried to use the time he had left to groom his officers to think like commanders. He looked for teachable moments. There were only about ten seats at the table, so after the ranking officers sat down, the others had to stand.

First up was the weekend's operations duty officer, who went over the past two days' events using a map of the region projected onto a screen. He pointed out search and rescue missions completed over the weekend, reports of hazardous conditions, and reviewed where all aircraft and personnel from the station were deployed at that moment. Next was the law enforcement duty officer, who recounted news from the airborne drug patrols, including sightings of suspicious vessels, arrests, and drug seizures.

The last briefing came from the station's hurricane officer, a position that was activated only during hurricane season. Lieutenant Jeff Henkel focused his remarks on the previous night's forecasts from the National Hurricane Center identifying the low-pressure system. He put the NHC's satellite pictures on a screen, alongside the forecasted track for the system. This was by far the most significant news Lorenzen took in that morning. He had been at the Clearwater base long enough to know not to underestimate a low-pressure system during hurricane season, even if it was forming in an odd spot. No South American drug trafficker or reckless weekend sailor posed as big a risk to life (and property) as a simmering low north of the shipping lanes on either side of the Bahamas.

Like the forecasters at the National Hurricane Center, Lorenzen had his people monitor weather maps and a multitude of information sources constantly for signs of trouble during hurricane season. As the captain liked to say, as soon as a puff of air wafts off the African coast, he wanted to know about it.

"Okay, what would you do if you were sitting in my chair?" he asked the room. "Where would you position the aircraft?"

Somebody mentioned that since the new hangar on Great Inagua was rated for Category 5 winds, it might be smart to leave the helicopters on base. Someone else disagreed, saying it would be wiser to bring aircraft back to Clearwater. *What about flying them to Gitmo?* a third guardsman asked, referring to Guantánamo Bay, Cuba.

Lorenzen listened to their input for a bit. He liked the Gitmo idea; it was good to have assets flanking a storm. He didn't like the idea of moving assets back to Clearwater. Too far from where they would be needed. Then he took the floor, standing in front of the weather maps.

Lorenzen pointed at 27.3 degrees north and 68.9 degrees west, where the low was swirling. "If this develops any further," he told the room, "we have to have assets ready. If it tracks to the north, this is where I anticipate moving our aircraft. If it goes westerly, I'll put them here." In each case, he wanted planes and helicopters behind the storm, ready to move quickly into damaged areas.

"We are the only game in town," he continued. "Smaller helicopters on ships won't have the range to conduct search and rescue missions, and if the seas turn rough, ships will be of little value for recovering people at sea." Instead, Lorenzen said, he wanted Jayhawk helicopters, which have a range of seven hundred nautical miles, abeam of the storm, to the left or right of it, and astern, or behind it. In a crisis situation, he liked to have helicopters following a storm from one hundred miles away, ready to pick up the pieces. "Everybody in the path of a hurricane is of no value" to help, he reminded the room.

After establishing the grand strategy, he adjourned the meeting, dismissing everyone except his command staff: his executive officer, operations officer, hurricane officer, aviation engineering officer, and the command enlisted adviser, his top enlisted man. This smaller group continued to develop their battle plan. The first order of business, everyone agreed, was to cancel the crew shift swaps at Great Inagua and Andros Islands. No sense conducting an operation that could risk equipment malfunctions, and it was best to keep the existing crews out there. They had been there more than a week now, so they were already in their rhythm and would be ready to move quickly if called upon.

On the table was a suggestion to "flow assets south," or get an additional helicopter and crew out to Andros, which housed only one Jayhawk, and bring in another crew for Great Inagua. For now, though, the only order given was to cancel the crew swap.

Before leaving the conference room, Lorenzen turned to his hurricane officer and told him, "I want to know the minute anything changes."

While Captain Lorenzen was contemplating disaster at his meeting in Clearwater, there was very little to trouble Captain Gelera's mind as he prepared to set sail down in Miami. He was aware of the tropical low but was not concerned; the forecast showed it moving in the opposite direction from where he was headed. The day was clear and sunny, with a mild wind. Gelera walked the length of his ship inspecting the lashings that secured his cargo. The holds were filled and their sliding metal covers shut. On the cargo deck, which comprised the front three-quarters of the ship, cars had been strapped down, their tires and axles chained to giant metal cleats. On top of the cars, the mattresses were stacked. Over the mattresses, the seamen had flung giant tarps that were cinched securely to the deck. Everything looked in order. At that point the *Minouche* carried about one thousand metric tons of general cargo, worth about $2 million. From the deck, Gelera looked down the length of the turbid brown river. Today, finally, they were scheduled to leave.

At about 10:00 A.M. that morning, two tugboats arrived to tow the ship down the river and out into the bay where the channel pilot would meet them. This was a required, and not inexpensive, part of the journey. State law (for foreign-flagged vessels) and federal law (for U.S.-flagged vessels) require that large ships hire local pilots to handle their arrivals and departures. It is too risky for captains with little or no knowledge of local waters to try to wend what is essentially a horizontal office building through river bends, sandbars, tidal flats, and other locality-specific hazards. With ships this size, an errant turn of the rudder could inflict a spectacular amount of damage to local docks and channels, not to mention the ship itself. A grounded ship could block a port for days or weeks. So it was better for everyone involved—ships' captains, owners, and dock owners—to have knowledgeable pilots guide the ships out to sea. This particular tow would cost the *Minouche*'s owners about $1,200, plus another $300 for the channel pilot.

The tugboats secured lines to the bow and stern of the *Minouche* and, because the ship was too big to turn around, towed it stern-first downriver. The *Minouche* was a "dead ship," its engines off, as they got under way, gliding down the river alongside the railroad tracks and warehouses of North River Drive, then past the forest of glass-and-steel office buildings downtown until they sailed out the mouth of the river, where the Tequesta tribe had built huts on raised platforms millennia ago, living off the bounty of seafood where the river met Biscayne Bay.

As the ship slipped into the bay, the Port of Miami on man-made Dodge Island hove into view. On the island's north side was the cruise ship terminal where Royal Caribbean, Carnival, Norwegian, and other passenger lines docked ships so massive they seemed to block out the horizon. On the south side of the island was a container ship terminal, with more than a dozen mechanized gantry cranes. In 2013, Miami installed "Super Post-Panamax" cranes, referring to their ability to handle the supersize container ships put into use after the expansion of the Panama Canal was completed in 2015. Giant ships sail under the cranes and wait to be loaded. Crane booms lift the containers off the wharf and onto the ships, stacking them like interlocking Lego pieces. The process takes hours now, not days. When the ships are unloaded, the containers are stacked neatly on top of one another until they can be placed on eighteen-wheelers, which roll through a brand-new billion-dollar tunnel to the highway.

Inside these metal rectangles are the goods that restock the shelves of our Walmarts, Best Buys, Targets, and Home Depots with cellphones, laptops, blue jeans, power saws, drywall, sheets for your bed, and mattresses that, unlike the ones strapped to the deck of the *Minouche,* have never been slept on before. Refrigerated containers, called reefers, carry fish, meat, and fresh produce. As the writer Rose George deftly summed up in the title of her book about the cargo ship industry, these floating warehouses carry "ninety percent of everything."

All of this was a world away from the *Minouche.* In essence, the *Minouche* still operated like ships from eighty years ago. In those days, stevedores on the docks loaded cargo into a ship's hold manually, one box, bundle, or crate at a time. The loading of a general cargo ship was a skilled affair. The weight needed to be distributed so that the ballast kept the ship's keel even. Loading a ship this way took a long time, weeks or more. It also exposed the cargo to theft and damage. Dockworkers might knock a hole in a crate just to see what was in it. Accounting for "loss" at the dock was a routine part of the bookkeeping.

Then, in the mid-1950s, an American trucking entrepreneur named Malcom McLean had an epiphany while watching his trucks being loaded by hand before being driven to a dock and unloaded, again by hand, onto a ship. It was so inefficient, McLean wondered if there was a better way. What if the back of the truck could simply be lifted onto the ship? Thus the modern shipping container was born. It took a while to catch on. Bigger ships were needed, along with deeper ports to accommodate them. But eventually the container McLean envisioned became the standard of the industry, in a very literal sense. A modern ship's capacity for storage is measured in TEUs, which stands for twenty-foot equivalent unit, the size of the standard shipping container. Eventually, standardizing these containers cut the cost of shipping goods by almost half.

The impact this had on global trade is hard to overstate. Suddenly it was cost-effective to make goods overseas, where labor was cheaper, and ship the finished product to markets in America or Europe, where people had the money to buy them. The efficiencies sometimes defied logic. In one famous example, it was cheaper to have cod caught off Scotland shipped to China to be filleted, then shipped back for sale in Scotland, than it was to process the fish locally.

If the docks of the Miami River resembled an older way of doing things, Port-de-Paix, where the *Minouche* was headed, was straight out of the nineteenth century. The port, on Haiti's northern coast, had no cranes, nor a wharf big enough to accommodate cargo ships. Instead, the ships dropped anchors fore and aft roughly five hundred yards offshore, where they were then met by smaller boats to offload goods. The ship's derricks lowered the cargo, bundle by bundle, into the smaller craft, many of them simply wooden rowboats.

This is true even for cars and trucks. After countless trips to Haiti, Gelera remains amazed at this maneuver: Two wooden rowboats would position themselves under the derricks as a vehicle was lowered in such a way that the front wheels would rest in the bow of one rowboat and the rear wheels nestle in the bow of the other. The boats would then be carefully rowed to shore. The *Minouche* might carry twenty cars a trip, all of which would be ferried to the wharf in this manner.

Miami is only about 740 nautical miles (nautical miles are about 15 percent longer than a regular mile) from Port-de-Paix, but that doesn't begin to measure the distance between the places. *Minouche* crew member Jules Cadet, who grew up fishing in Haiti with nets from wooden skiffs, crossed this chasm regularly. On one day he'd sail past the Port of Miami's giant cranes, a riot of bright lights and humming machinery, an economy measured in millions and billions, and a few days and a few hundred miles later he'd sail into a harbor where rowboats met and unloaded ships in a way that was as primitive as it was ingenious.

Gelera didn't mind the simpler world in which the *Minouche* operated. In fact he appreciated it. He had captained other ships and worked at bigger companies. He liked the *Minouche*'s owner, Milfort Sanon, a Haitian businessman based in Miami, and the ship's agent, Richard Dubin, who was Gelera's main point of contact. They paid on time, and unlike at the bigger companies, Gelera was left alone to do his job. There was no meddlesome hierarchy over him, no land-based administrators second-guessing his calls or telling him what to do. So he'd take an older, less sophisticated ship for this degree of freedom. He still earned enough to put his kids back in Guatemala through college, including medical school for his eldest daughter.

And in truth, Gelera didn't even mind the ship itself. She was old but he liked her lines, her low profile, where the bulwark swooped down between the bow and the bridge. He thought it made the ship more stable. There was some rust, but the metal was good, which was important because she had to be able to take the waves—not just the pounding, but also have enough give to flex slightly in heavy seas. There's no denying that she'd seen better days. It had been thirty-five years since the *Minouche* rolled out of the shipyard and into the fjord in Ringkøbing, Denmark, back in 1980. She was 70 meters by 10 meters, or roughly 230 feet by 30 feet. She'd sailed the Mediterranean, the Baltic Sea, and the North Sea as a bulk cargo carrier for years. Then she'd crossed the Atlantic to run routes in the friendlier waters of the Gulf of Mexico and the Caribbean.

Gelera had actually visited the ship once, back in the early eighties, in one of those nautical coincidences that happen sometimes when you sail the world for a living. He was captain of another vessel, the *Sandakan,* running rice from the Port of Lake Charles, Louisiana, to Petit-Goâve, Haiti. On one trip to Petit-Goâve, he anchored the *Sandakan* offshore next to a new green-and-white Danish vessel named the *Lady Lotmore II.* One afternoon a member of the *Lotmore II*'s crew took a small boat over to the *Sandakan* to invite Gelera to a party they were having aboard the Danish ship. He accepted and went over for a few hours. He remembers thinking, *This is a very shiny ship. It smells like roses.* He drank some beer with the crew, mostly Filipinos and two Danish officers, and ate their grilled pork, then returned to his own ship. At the time, the *Lady Lotmore II* was only three years old.

Decades later, the ship ended up docked in Miami, no longer smelling like roses. She sat idle until Milfort Sanon bought her, presumably for a bargain. She was cleaned and geared up, and rechristened the *Minouche* in 2014. Gelera was the second captain to sail her under that name.

After she returned to service, the Coast Guard inspected her annually. Inspectors cited the ship for various housekeeping items, such as not having the ship's engineer present during an inspection and having the wrong name on a safety radio certificate. They found more serious issues, too, though nothing out of the ordinary for a thirty-five-year-old vessel. Railings and bulwarks on the port side had rusted through or were missing. A cement patch was found on a water discharge pipe in the ship's machinery space. If that patch broke, it could flood. But the ship's management resolved each issue, welding new metal onto the port rails and cropping and refitting the patched pipe. In the end, the ship was always found fit to sail.

By now she was painted red, black, and white and sported a cargo mast and two booms for loading and unloading cargo. It took a crew of twelve to run her. And while she called the Miami River home, she sailed under a Bolivian flag.

It's a safe bet that she never sailed to Bolivia, which is landlocked.

In this, the owners of the *Minouche* were following a long-established maritime tradition—flying a flag of convenience. This meant the ship did not have to abide by U.S. labor laws, even though it operated out of Miami. Registering the ship in Bolivia was the equivalent of a New Jersey company having a PO box in Delaware to avoid paying corporate taxes in its actual home state.

In the world of maritime trade, this is the norm. According to UN numbers, of the 1,995 U.S.-owned merchant ships registered in 2015, 1,213 flew foreign flags. American shipping companies are not alone in this. The United Kingdom had a merchant fleet of 1,329 ships; 997 were foreign flagged. Of Japan's fleet of 3,969 ships, a whopping 3,134 flew a foreign flag. On and on and on. All told, some 70 percent of the world's commercial fleet are foreign flagged.

The reason is simple: Filipino, Indian, Chinese, and even Russian crews cost far less than American, British, and Japanese ones. Flying a foreign flag can cut shipping costs drastically—a Third World crew member will earn a quarter to half of what a First World sailor makes. Not to mention compliance with costly regulations that the foreign registry would never—could never—impose.

The UN Conference on Trade and Development stated pretty plainly in a 2016 report that this massive loophole is the primary reason—arguably the entire reason—that ship owners in developed countries have been able to operate profitably in the global marketplace. It "is due to the system of open registries that they [developed countries] may remain competitive against fleets owned by companies based in developing countries. For example, under the flags of Liberia, the Marshall Islands or Panama, an owner from Germany or Japan can employ third-country seafarers, for example from Indonesia or the Philippines, who work for lower wages than their German or Japanese colleagues."

Ships have been using foreign flags to get around regulations for centuries. But in the modern era the practice began to spread in the 1920s, when Panama opened its ship registry to foreign vessels. The first foreign vessel to register in Panama was a rum-running ship during Prohibition that moved booze from Canada to the United States. (Because the ship was considered to be a floating piece of Panama, U.S. laws didn't apply.) A few years later, two U.S. cruise ships flew the Panamanian flag so they could serve alcohol to passengers once they were in international waters.

Soon, open registries began to make sense to ship owners as a way to cut labor costs and reduce federal oversight. In the 1940s, U.S. businessmen collaborated with the government of Liberia to create an open registry that could compete with Panama's. Today, Panama, Liberia, and the Marshall Islands have the largest ship registries in the world: 41 percent of world tonnage.

Yet for all this history of shaving costs—finding the cheapest labor, dodging regulations, maximizing the amount of cargo a ship can carry, and trimming the time and labor needed to load it—it hasn't been enough. After barely surviving the recession of 2008, many shipping companies ordered more ships to try to recover lost earnings. Driving these orders was the renovation of the Panama Canal, allowing bigger ships to pass through. But the firms overestimated the demand and ended up with too much cargo space and not enough cargo, especially after China's economy cooled. By 2015, nearly every segment of the shipping world, except for oil tankers, had suffered historically low freight rates and diminished earnings, the result of low demand and too many ships. Now many of those new ships sat idle, or worse.

Faced with the incredibly high cost of running a cargo ship— fuel, salaries, and docking fees alone can run $10,000 to $20,000 a day—some firms were turning their ships into scrap metal. In 2016, the Singaporean firm Rickmers Maritime Trust sent a seven-year-old cargo ship, the youngest ever, to the scrapyard. The owners simply couldn't afford to keep it afloat. That same year, after one of the world's largest shipping firms, South Korea's Hanjin Shipping Company, declared bankruptcy, ports refused to allow its ships to dock and unload, assuming they wouldn't get paid if they let the ships in. Hanjin's ships were marooned at sea with $14 billion worth of cargo on board, until a series of court decisions gave the vessels protection to unload in the United States.

Say what you will about the trade in secondhand mattresses and a cargo terminal that relies on rowboats, but it was steady work, safe from the vicissitudes of globalization.

Then there is the weather to worry about. Global warming may not pose as immediate a threat as falling freight rates and bankruptcies, but its impact will be much longer lasting. One study by noted MIT climate scientist Kerry Emanuel has forecast that the future may have fewer storms, but stronger ones—more Category 4 and 5 hurricanes.

For reasons not yet understood, but possibly related to the increase in total water volume from melting polar ice caps and thermal expansion, extreme wave heights during these hurricanes may also increase. Scientists have been recording increasing wave heights over the past several decades in various places around the world—and expect that trend to continue and in some cases accelerate. A team of American researchers recorded an average of more than half an inch of "deep water significant" wave-height increase per year in the Pacific Northwest since the mid-1970s. Canadian scientists, using climate change models, projected that extreme wave heights are likely to double or triple in several coastal regions around the world, including the tropics, over the next fifty years. Climate change models developed by Japan's Disaster Prevention Research Institute projected a wave height increase in the middle latitudes, and a decrease around the Sea of Japan. Meanwhile, "extreme waves due to tropical cyclones will be increased" in all regions.

Complicating the science of detecting trends in hurricanes is that there are naturally occurring climate cycles that impact hurricane activity. The most well-known of these cycles, El Niño, occurs roughly every two to seven years, when warm water in the tropical Pacific shifts eastward. Another such cycle, the multidecadal oscillation, creates periods of increased hurricane intensity in the Atlantic. While there is growing consensus among scientists that climate change will shape the behavior of hurricanes, these other natural cycles complicate the prediction of future activity.

Once the tugboats had guided the *Minouche* into Biscayne Bay to the Dodge Island Channel, the channel pilot climbed aboard, using a pilot ladder, to navigate the ship out of the bay, through Government Cut—a narrow passage with wealthy Fisher Island, on one side and garish South Beach on the other—and into the open Atlantic. The channel pilot's job was to sail vessels along lanes cut for the big ships. To stray from these channels meant running aground.

After arriving at the Miami sea buoy, the channel pilot handed the bridge back to Captain Gelera, climbed down into his motorboat, and sped back toward the city. Gelera, finally back in control of his ship, set his course for the next waypoint, Ocean Cay, about thirty-six miles southeast. From the deck, the Atlantic lay flat in the sunshine like a mirror on the floor. There was some light air, less than five miles per hour, blowing out of the south-southeast. In fact, there was so little weather that Gelera didn't bother to record it in his logbook. A day like this gave rise to the cliché "smooth sailing," the kind of day the captain of a cargo ship prayed for. No weather meant a faster cruising speed and a lower fuel-burn rate. Gelera told his men they were free to throw some fishing lines overboard, and in short order they had landed about a half dozen mahi-mahi. But the crew would not eat the fish fresh that night. Instead, they salted and dried the fillets to bring home.

As the late afternoon bled into dusk and the sun began its descent into the horizon, Gelera checked the NHC's latest forecasts. The five o'clock report indicated that Tropical Depression Eleven was still swirling about 465 miles southwest of Bermuda, heading west, away from the *Minouche,* at six miles per hour. Unconcerned, the captain headed to the galley for some chicken and rice, happy to be under way.

But over the next few hours, the storm strengthened ever so slightly. The winds increased from 30 to 40 miles per hour. And, strangely, it changed direction, pivoting southwest at five miles per hour. The uptick in wind speed was enough to earn the weather system a modest promotion by the NHC. At 11:00 P.M., as Gelera slept in his cabin, Tropical Depression Eleven graduated. It was now Tropical Storm Joaquin.

Chapter 4
PORT OF JACKSONVILLE

The starboard side of the SS *El Faro*'s hull, a towering wall of blue-painted steel, loomed over the wharf at the Port of Jacksonville's Blount Island terminal as gantry cranes loaded her decks with cargo containers. She was a steamship (designated by the "SS" before her name), using two large boilers to power a single-propeller shaft. And she was old, built by the Sun Shipbuilding and Dry Dock Company in Chester, Pennsylvania, just south of Philadelphia, which rolled her into the Delaware River for service on January 1, 1975. That put her in the minority of big ships—less than 9 percent of the world's merchant fleet is over twenty years old. In fact, steamships are mostly a relic from an earlier era, replaced by the diesel- and natural-gas-powered vessels that are much more common today.

El Faro showed her age. Rust streaks stained her from bow to stern, and her deck was downright orange with oxidation. But she was still impressive: 17,527 tons of welded steel stretching 790 feet, almost three football fields long, 92 feet across, with a towering multi-story white superstructure housing the bridge, crew quarters, and galley. In the depths of the ship was a pair of immense boilers that could produce thirty thousand brake horse-power for a maximum speed of 24 knots. Below the main deck were three more decks, divided up into five massive cargo holds, including a semi-enclosed deck with ramps so vehicles could be easily rolled on and off the ship.

El Faro was one of ten similar ships built between 1967 and 1977 that were grouped together as the Ponce de León, or Ponce for short, class of ships. She was originally christened the *Puerto Rico* and assigned to trade routes in the Caribbean and Gulf of Mexico. Twenty years later, in the mid-1990s, she was hauled into a shipyard in Alabama, where her mid-body was lengthened to increase cargo capacity. Rechristened the *Northern Lights,* she spent the next ten years or so sailing cargo to ports in Washington State and Alaska. In 2006 she was hauled out of the water again for some deck conversions and to have nearly five thousand pounds of iron ore fixed to the ballast tanks to stabilize the now bigger ship. She received one last name change, to *El Faro*—Spanish for "lighthouse"—and was moved to Jacksonville for the Puerto Rico route. (The ship's "hailing port" was actually San Juan.) *El Faro*'s parent company, TOTE Maritime, which operates ships with routes to Puerto Rico and Alaska, and TOTE Services, which manages the personnel on those ships, are based in Jacksonville. Both are subsidiaries of TOTE Inc., headquartered in New Jersey.

Sun Shipbuilding has since closed, a casualty of America's decline in manufacturing, leaving a dwindling number of shipyards able to construct big cargo ships in the United States, which also means a dwindling number of shipyards capable of fulfilling the requirements of the Jones Act, a 1920 law requiring that any cargo transported from one U.S. port to another must travel on ships that are American built, American crewed, and American owned. (Puerto Rico, being a U.S. territory, counts as a U.S. port.)

The law was designed to protect America's supply routes during times of war. Today its primary effect is to protect the jobs of American sailors, preventing companies from hiring much cheaper crews from Third World countries. But there is a cost—and it is steep. To build a Jones Act ship costs $120 million to $140 million. To build the same ship in South Korea, which is a developed nation, would cost about $32 million, according to Court Smith, an industry analyst with Shipping Intelligence and Analytics. It's even cheaper to build one in India or China. South Korea builds roughly two hundred commercial ships a year, according to Smith. America puts out maybe four.

As a result, shipping companies pushed the life spans of their expensive American-made ships to the absolute limit. The average age of the U.S.-flagged cargo fleet is thirty-three years, compared to thirteen years for the global fleet, according to UN statistics, and most shipping experts say the average age a cargo ship is retired worldwide is around twenty years. *El Faro* was a product of this dynamic. Due to its age, it was allowed to remain outdated in certain areas. For example, a regulation requiring new ships to carry enclosed lifeboats was waived for older ones, for which compliance would require a costly retrofit. Grandfathered in, *El Faro* continued to carry two old-fashioned open-top lifeboats. Likewise, the ship's emergency position-indicating radio beacon, or EPIRB, did not have to be encoded with GPS, which would give the ship's position in a time of distress.

Still, the ship passed a Coast Guard annual oversight inspection in March 2015 and was deemed fit for another year of duty. But *El Faro*'s inspection process was about to become more stringent. Inspectors with the Coast Guard's Alternate Compliance Program put her on its targeted-vessel list for 2016, one of the roughly 10 percent of commercial ships that, due to age, ac-

cident history, or ship type, were deemed to be at heightened risk
of marine casualty.

El Faro had just returned from a run to Puerto Rico on Monday,
September 28. She arrived at the Blount Island terminal at 12:42
P.M. for loading and a quick turnaround. Stevedores and long-
shoremen immediately unloaded inbound cargo and then began
loading the outbound cargo. Gantry cranes lifted containers
onto the ship's decks, while cars and wheeled cargo containers
were rolled up ramps to the lower decks. By Tuesday evening she
had 391 cargo containers on her main deck and 149 automo-
biles, 118 trailers, and 238 refrigerated containers belowdecks.
Lashing gangs secured the cargo with chains and ties. Meanwhile, the ship's crew prepared for an 8:00 P.M. departure for San
Juan. Turnaround was that fast.

The weight of the containers and cars was entered into a stability and load management software program called CargoMax,
which helped determine weight distribution, taking into account
how much fuel was loaded and how much would burn off during
the journey. This was to ensure that the ship was safely balanced
and maintained enough reserve buoyancy. As the containers
stacked up, three high, on *El Faro* Tuesday, the ship developed a
starboard list of about four degrees, so the terminal manager
ordered cargo to be loaded to the port side, correcting the list.

Following the financial collapse of one of TOTE's competitors on the Puerto Rico route, Horizon Lines, a year earlier,
TOTE was taking on more cargo than it had before, including a
sizable spike in refrigerated containers. During the previous ten
years, a big load on the two TOTE ships that sailed to Puerto
Rico, *El Faro* and *El Yunque*, would have been ten thousand
tons of cargo, according to the later testimony of one of TOTE's
captains, Earl Loftfield. After Horizon disappeared, TOTE's
ships would frequently sail with twelve thousand tons. "We were

loading as much cargo as we could," Loftfield said. To do this, the officers in charge pushed load limits to within inches of the allowable "metacentric height limits." Metacentric height is a measurement of the ship's stability. Exceeding the limit can destabilize a ship.

There were thirty-three people aboard *El Faro* for this run to Puerto Rico: a crew of twenty-seven American sailors; five Polish welders working to refit the ship for its next assignment in Alaska; and a TOTE engineer whose job was to oversee the Poles.

Many of the sailors were at the tail end of their contracted time, having come on during July and August for a month or two of duty. One exception was third assistant engineer Dylan Meklin, a robust blue-eyed boy from Rockland, Maine. A standout high school athlete who'd excelled in football, basketball, and baseball, Meklin had graduated from Maine Maritime Academy that spring and become a member of the American Maritime Officers union. He was a last-minute hire for this trip, which would be the twenty-three-year-old's first-ever job at sea. It didn't start well. TOTE Services misspelled his name on the itinerary they sent him and had to reschedule his flight. Then his plane was delayed over Boston, causing him to miss his connecting flight. By the time Meklin landed in Jacksonville he was a nervous wreck, racing from the airport to the ship.

"He was frantically texting me," recalled his mother, Elaine Meklin, "because he needed his check information for his direct deposit." He finally made it to the ship at about 7:00 P.M., an hour before *El Faro* was scheduled to depart. "He called just to say he was on the ship and he got the information, just a short conversation," she said. "Oh, and to tell me he knew someone from Rockland was on the trip, which was Danielle [Randolph, the ship's second mate]." The two hadn't met before, but they were from the same town and the same college. In fact, there were so many alumni from his school on board—five of the officers—they were called the Maine Maritime Mafia.

At about 7:30 P.M., river pilot Eric Bryson boarded the ship and made his way to the bridge. *El Faro*'s third mate, Jeremie Riehm, a handsome forty-six-year-old from Florida, with a mane of brown hair that shaved a good ten years off his appearance, greeted Bryson by pointing to a stand by the wall. *The coffee's fresh*, Riehm offered. Bryson filled a mug. He had piloted this ship dozens of times and was familiar with the bridge. He checked his pilot's card for *El Faro* (he had one for each of the ships he regularly piloted), which had notes on electronics, gear, and draft. *How's she running?* he asked. *Everything's fine*, Riehm told him.

Outside, on the water below, tugboats secured lines to the bow and stern to guide the ship away from the wharf and into the middle of the St. Johns River. There the tugs would turn the ship around, pointing her bow east toward the Atlantic. Bryson would oversee the roughly nine-mile journey to the mouth of the St. Johns, where it opens into the Atlantic Ocean.

As Bryson was sipping his coffee, the quartermaster, Jack Jackson, a sixty-year-old able seaman, whose sailing career had taken him from Africa to the polar ice caps, arrived on the bridge. *Howdy, Captain*, Jackson said. The two men had worked together plenty of times before. Jackson would be at the helm while Bryson was in charge. A few minutes later, *El Faro*'s master, Captain Michael Davidson, joined them.

A lean, clean-shaven fifty-three-year-old with neatly trimmed silver hair, Davidson had the well-groomed air of a company executive more than that of a sailor. In fact, he demanded that his officers also be presentable, chiding them to shave and cut their hair. But he was definitely a sailor. He had been working boats his whole life, ever since he was a kid in Maine, and had spent the early years of his career sailing on oil tankers and cargo ships in the rough waters off Alaska.

On the bridge, the men discussed the tides, the draft of the vessel, and what would have to be done to maneuver the ship safely down the wide and roiling river.

Mariners have been using the St. Johns as a trade route for

centuries, giving it a storied place in American merchant marine history. In 1565, English traders sailed up her waters from the Atlantic to a French encampment named Fort Caroline. The Englishmen traded a boat and some food to the French in exchange for ammunition and weapons. Someone scribbled the transaction down, and it became the first record of international maritime trade in America.

Because of that boat-for-weapons swap with the French, Jaxport's boosters like to call it America's First Port. In reality it's one of the newer ports, officially coming into existence in 1963.

And while *El Faro* had been operating out of Jaxport since 2006, the TOTE Maritime name was new that year. The ships were originally run by the Sea Star Line, of which TOTE Inc. had become a co-owner back in 1998. The 2015 name change was part of a rebranding effort to distance itself from a price-fixing scandal that had engulfed the company in 2008.

For two decades, Sea Star had been one of several shipping companies that sailed supplies to Puerto Rico on a regular schedule. All that competition meant slender margins and low profits. But companies stayed with the route to Puerto Rico—travel between a U.S. state and a U.S. territory—because they had invested so much to comply with the Jones Act. There were no dodgy flags of convenience for *El Faro;* she flew the Stars and Stripes. That was a high barrier to entry for would-be competitors—they would need an American-built ship, crewed by U.S. sailors.

Then, in 2002, one of those firms, Navieras, went bankrupt. Suddenly the companies still afloat—including Sea Star, Crowley, Trailer Bridge, and Horizon—started seeing increased business and profits for the first time in decades. Rather than risk losing their newfound earnings to any potential newcomers, the companies bought Navieras's ships, and executives from at least three of the companies, Sea Star, Crowley, and Horizon, conspired to fix their prices. They created secret email accounts to communicate, and set up spreadsheets that kept track of their rates, which increased by as much as 30 percent—so that, as *Forbes* magazine wrote, "they could assure that nobody was cheating, while they were cheating." They weren't clever enough to fool the FBI, however, which got involved after learning of a meeting between executives from the competing firms.

As much as the collusion that followed was a symptom of greed and a high degree of ethical flexibility, it was also emblematic of the desperation that often runs through an industry defined by narrow margins and huge capital outlays. Eventually six people, including Sea Star's president, Frank Peake, would be sentenced to prison, and the companies would pay multimillion-dollar fines. Crowley Liner Services pleaded guilty to price-fixing and was fined $17 million. In the aftermath, Horizon had to refinance to avoid bankruptcy; it limped along for two more years until it could no longer survive and was sold. The case continued to drag on. Even as *El Faro* prepared to cast off that Tuesday night, Sea Star's convicted president was appealing his sentence.

In the aftermath of the scandal, Sea Star changed its name to TOTE Maritime and continued sailing, making the officers, crewmen, and unions happy. Manipulating cargo rates and conspiring with competitors was the work of accountants and desk riders, "the pencil pushers who owned the world," as one *El Faro* crew member put it. It had little to do with sailors on ships. All the merchant mariners really cared about was that TOTE was one of a handful of companies offering good-paying jobs to American sailors. And TOTE showed optimism about the future of the route. Two brand-new liquefied natural gas–fueled ships the company had commissioned were being delivered to Jacksonville that year.

By about 8:00 P.M. the tugs had towed *El Faro* away from the wharf and maneuvered her bow east toward the mouth of the St. Johns, and just before 8:30, the ship cast off the lines to the tugboats and Bryson took control of the vessel, giving steering commands to helmsman Jackson.

It was a calm night under a full moon, its pale light shimmering on the water. The journey downriver took a little over an hour. They passed one other cargo ship, the 637-foot CMA CGM *Kingfish,* sailing upriver. As they steamed east, Bryson made small talk with Davidson. They chatted about the weather, and Tropical Storm Joaquin came up. *If it looks like trouble,* Davidson told Bryson, *I'm going to shoot down under it.* This echoed what he had told an off-duty second mate named Charles Baird, who had texted the captain from his home in Maine that morning when he saw the weather reports. The night was calm enough and the storm was far enough away that Bryson didn't give it another thought. "Very typical, it was a routine trip out," Bryson recalled. "The captain was a pro, very much into the job and doing it correctly."

Sometime around 9:30 P.M., after sailing past the jetty where the St. Johns met the Atlantic, they made the last buoy in the shipping lanes. Bryson's work was done. He said goodbye to the captain, the third mate, and the helmsman, wished them a good trip, and climbed down a ladder off the lee side of the ship to the waiting pilot's boat that had been following.

Davidson then took back control of his ship and gave the order to turn *El Faro*'s bow almost directly south-southeast, following a track line that would keep her on the Atlantic side of the Bahamas until they reached San Juan, Puerto Rico, on Thursday, 1,400 nautical miles away.

A career merchant mariner, Davidson crewed on Casco Bay ferries during his teenage summers before attending Maine Maritime Academy, where he graduated in 1988. He started work in Alaska the next year, which happened to be the year when the oil tanker *Exxon Valdez* ran aground, spilling more than ten million gallons of crude oil, an environmental disaster that would ruin the reputation of its captain, Joseph Hazelwood. For more than a decade Davidson toiled in Alaska and Washington for oil companies such as Texaco and ARCO, working his way up to chief mate. The sailors he worked with described him as "meticulous."

He was also ambitious, always on the lookout for other opportunities. While in Alaska he studied to get his pilot's certificate for the bays and inlets around Valdez. These are some of the most hazardous waters along America's shoreline, shot through with powerful crosscurrents and pocked with dangerous shoals. Winter storms can be brutal, with high winds and crushing waves. Earning a pilot's license there took steady nerves and technical mastery.

In the end, he stuck with the open water, leaving the Northeast to captain a research ship from Bermuda in 2004, then joining various shipping firms, such as Sabine, InterOcean (a TOTE company), and Crowley, where his employment there didn't end well. He received two letters of warning, one regarding an accident report, the other for failure to report cargo damage. At one point, Davidson refused to take a ship from one port to another after a team of engineers inspected the steering mechanism and told him it was unsafe. "The people in the office wanted him to do it anyway," his wife, Theresa, recounted. "And he said, 'I'm not going to do that.' So he ordered two tugs to move the ship and when he came back from vacation they weren't too happy with the bill and told him he was no longer employed." Her husband, she added, was the kind of man who wanted to do everything by the book.

With two young daughters, Davidson was desperate to get back to work, so in 2013 he took a third mate position with TOTE—a step down, but it was a job. It wasn't long before the company recognized the depth of his marine experience and education. TOTE sent him to Jacksonville to help with a personnel shake-up after the company parted ways with two captains. One of those captains, who piloted *El Morro*, a sister ship of *El Faro*, had been fired. The captain alleged it was because he had complained to TOTE management about safety issues. TOTE disputed this, saying that the company lost confidence in the captain after members of his crew were caught smuggling forty-three kilos of cocaine aboard the ship. In any case, Davidson's job was to clean up the mess. His first command with TOTE was *El Morro*. But in 2014, shortly after he took the helm, the ship was consigned to the scrapyard. She was forty years old at the time, her steel too far gone. TOTE then assigned Davidson to *El Faro*. She was only a year younger, so everyone assumed it was only a matter of time before *El Faro* would be scrapped as well.

Davidson believed he was working hard for the company, taking on these older ships without complaint. His goal was to get assigned to one of the two new natural gas–fueled ships the company had commissioned. They would be the very first such container ships in the country. He had put in for one position that was open, but a few months before his August contract started TOTE's crew manager told him he hadn't been selected. Instead, TOTE offered him a job taking *El Faro* up to Alaska.

Davidson was hugely disappointed. He wondered if this meant his time at the company was about to run out. TOTE apparently wanted younger crews on the new ships, sailors who were more comfortable monitoring the ship's functions on computer screens and using joystick-controlled thrusters to manage its position. Davidson began looking for other jobs. But he also seemed to find peace with his situation. His wife recalled an email he wrote her stating that he was "just going to keep doing what I do, the best I can, and see what happens."

All three of TOTE's Ponce-class ships were old and rusty. Claudia Shultz, the wife of *El Faro*'s chief mate, Steve Shultz, remembers getting a tour of *El Yunque*, the ship her husband worked on before he was assigned to *El Faro*. There were rust spots everywhere. Crew members used grinders to remove the rust and repaint the spots, but some areas seemed beyond repair. "Better look out for that one—it looks like you'll fall right through it," Claudia warned her husband as they walked around the bridge deck. "We all know about it," he replied. "It'll get taken care of."

El Faro was just as bad.

Kurt Bruer, an able seaman who sailed on *El Faro* in February 2015, recalled that when they were doing maintenance on the deck they would fill whole barrels with the rust they had scraped off. "Everywhere you turned there was rust," he said. "There was even rust in my room."

Danielle Randolph, *El Faro*'s thirty-four-year-old second mate (and a rare female merchant mariner), horrified her friends back home in Maine with tales of the decrepit ship. She told them *El Faro* was wasting away beneath her feet, regaling them with stories about scraping the ship while doing maintenance and coming away with huge chunks of rusty metal. She told them about finding bolts so rusty she didn't know how they were holding things together. She talked about how frustrated she was with the condition of the vessel, and that it would have to be scrapped soon. About a day before sailing from Jacksonville, Randolph sent a friend a photo of some graffiti on the boat. "God Be With Us All," it read. *Apparently I'm not the only that's having a hard time up here,* Randolph wrote.

She wasn't.

On September 8, 2015, just before Jack Jackson, the veteran able seaman, was scheduled to ship out on *El Faro,* he told his brother he didn't feel good about the "rust bucket."

"Every other Tuesday we talked for, like, five, six, seven minutes," Glen Jackson recalled. "But this Tuesday we spoke for forty-three minutes. He was apprehensive. For the first time ever, he told me he was thinking of breaking a contract and getting off the ship. He wasn't confident in the ship."

The condition of the steel wasn't the only thing that made

people nervous. *El Faro*'s age-based exemptions to certain safety regulations did, too—especially the one pertaining to its lifeboats. Because the ship was built before 1986, the year new regulations went into effect requiring enclosed lifeboats for ships of its size, *El Faro* did not have to abide by the rules. So the ship had old-fashioned open-top lifeboats mounted on gravity davits, small cranes used to raise and lower them. Randolph never complained to her mom, Laurie Bobillot, about the rust, because she didn't want her to worry (and because her mother had raised her not to be a complainer). But she did send her a picture of one of *El Faro*'s lifeboats. Bobillot was shocked. She had served in the Navy and raised her kid on the water in Maine.

"That is your lifeboat?" she asked, incredulous. "It's open." Randolph replied, "Yes, it's open." Her mother shot back: "Let's hope you never get into some rough seas because you know, kid, you're screwed." (The ship also had five inflatable covered life rafts on board that could hold a total of 106 people, but those could only float; they could not be steered and propelled.)

Randolph could take the tough talk from her mom. She was five foot three inches tall—"five foot nothing," her mom would tease her—and cheery, with blue eyes, blond hair, and a seemingly perpetual smile. Yet she was as strong as new steel (not the rusty crap she was scraping off the deck of *El Faro*). She was often the only female on board, though on this trip she was one of two women crewing *El Faro*. The other was Mariette Wright, a fifty-one-year-old utility deck worker from St. Augustine, Florida.

For Randolph, a career at sea was her life's ambition. "She learned to row a dingy when she was about five years old," her mother said. "That kid has been on the water as far back as I can remember."

When it was time for college, Randolph applied to Maine Maritime Academy—and only Maine Maritime Academy. Her mother encouraged her to apply to other schools as well, just in case. "But she was headstrong," Bobillot said. "It was full-bore ahead." Maine Maritime accepted her, and she graduated in the class of 2004, working her way up the ranks to second mate. She'd be a captain and have her own ship someday, she told her mom. If she had to sail on a creaky old ship for a few years to get there, so be it.

Keith Griffin, *El Faro*'s round-faced thirty-three-year-old first assistant engineer, had always enjoyed working on the ship's antiquated steam engine, at least according to his brother, Richard. Keith liked a challenge, and he was stubborn. The Griffins were, by nature, stubborn. They were New Englanders, after all, raised in Winthrop, Massachusetts, a spit of land jutting into the water east of Boston's Logan International Airport. The rock stars of the neighborhood when the boys were growing up were the men who worked the sea: lobstermen, fishermen, and merchant mariners. These were tough guys who went out in all kinds of weather. Griffin's father worked for the phone company, but he was an avid fisherman and took his kids out on the water every weekend. Both boys, not surprisingly, attended Massachusetts Maritime Academy. Keith became a ship's engineer. Richard stayed on the deck and joined the Coast Guard reserve. As Keith set sail on *El Faro* that day in late September, Richard was getting ready to deploy to Guantánamo Bay, Cuba, as a boat chief providing port security.

Sometimes their paths crossed during work. About a month earlier, Richard had been in Puerto Rico just as *El Faro* was docking, giving him the rare chance to have lunch with his brother. The two men met over fish tacos. But before they could even finish their meals, Keith got a call on his cellphone. *Gotta go*, he told his brother. *Something needs fixing in the engine room. They need me.* Keith wasn't even upset, Richard recalled. He seemed excited.

But as the tour wore on, Keith's patience was thinning. He had other things on his mind. His wife of two years, Katie, was pregnant with twins back in Fort Myers, Florida. They didn't know whether they were having sons, daughters, or one of each. They were waiting for Keith's tour to end on October 13 so that they could find out together. And despite the brave face he put on for his brother, Keith told Katie that he was ready to get off the ship. It was just constant work.

Some of the relentless pace undoubtedly came from Griffin's supervisor, thirty-four-year-old chief engineer Richard Pusatere, who was punctilious and wholly focused on work. "He went to work and he stayed at work," his father, Frank Pusatere, said about his son. "He just delved into everything one hundred percent." Griffin would gripe about his boss's OCD, but always jokingly, Katie Griffin said. Pusatere trained Griffin for his first job at TOTE, she said, and the two men had always gotten along well.

Pusatere was more circumspect about the state of *El Faro*. If he didn't like working on an older ship or had concerns about its seaworthiness, he didn't bring it up with parents or his wife. He did, however, tell a friend back home in Maine that the crew the company had hired to refit the ship to prepare it for the Alaska trade were the lowest bidders—by hundreds of thousands of dollars—and that they were wasting enormous amounts of his time asking how to put the parts they were working on in the engine room back together. "He said, 'Finally I had to go to my supervisor and tell them—Look, I can rebuild the boilers but I've got my own job to do. They're going to have to do it themselves,'" the friend, Tim Grunwald, recounted.

That alleged low bidder was Intec Maritime Offshore Service Corporation, a European company with offices in Switzerland, Germany, and Poland, originally founded to work on oil rigs in the North Sea. Later, the company began servicing cruise ships and opened an office in Fort Lauderdale, Florida. Intec had five workers on board *El Faro* for the September 29 journey, the Polish welders. They were there to replace the aging ship's winches, including the cable trays, in preparation for *El Faro*'s pending reassignment to Alaska.

For all the complaints about rust, *El Faro*'s crew may have been consoled by the fact that the ship had recently passed one of the routine inspections meant to catch vital safety issues. There was a time when these inspections were done solely by the Coast Guard, but after lobbying by the shipping industry, the service agreed to a third-party-compliance program in the early 1990s. Ship owners could now contract certified trade groups, such as the American Bureau of Shipping (ABS), to conduct inspections for them. The Alternate Compliance Program, as it was called, eased the burden of an overextended Coast Guard and gave ship owners more flexibility. But it also posed a potential conflict of interest. After all, it was the ship owners who hired the ABS inspectors in the first place—and paid them for their work.

"The Alternate Compliance Program (ACP) is a failure," Captain John Loftus, a shipmaster at Horizon Shipping before it went bankrupt, wrote in an open letter on the marine website gCaptain.com. "They are not thorough, and the USCG, with reduced manpower, has been reduced to do a quick walk around the vessel, and accepting the ABS Inspector's report. It is not a good system."

Loftus, who sailed the Jacksonville–to–San Juan route for Horizon, is outspoken on ship safety. In 2013, Horizon abruptly fired him after he reported safety violations on his ship to the ABS and the Coast Guard. In response, Loftus invoked the Seaman's Protection Act, which was designed to protect mariners in these situations. He sued and won a judgment in excess of $1 million.

Still, potential conflicts of interest didn't negate the fact that, for the most part, the U.S. shipping industry was well regulated, with plenty of rules meant to protect the lives of its sailors. For all the concerns some of the crew had about the age of the ship and its rust, they most likely believed that the system—TOTE, the inspectors, and the Coast Guard—would not let the ship sail if she was not seaworthy. So as *El Faro* steamed south Tuesday night, it's safe to assume that everyone but the shift's watch were resting easily in their cabins, swaying on the light seas below them: thirty-one men and two women with lives still unfolding, from new employees just starting their careers to veteran sailors wondering how they'd fit into shipping's changing world.

Nearly four hundred miles away, a ridge of high-pressure air over the Western Atlantic was blocking Tropical Storm Joaquin from moving north as forecast. Instead it began to move southwest-ward. The storm had slipped down and was now hovering over the Bahamas. Meanwhile, the vertical shear that forecasters had thought would weaken the storm, instead itself weakened. The center of the system was in an area of deep convection, meaning it was surrounded by and embedded with thunderstorms. The NHC forecasters warned that Joaquin was now over warm water "conducive to intensification." The storm was feeding on the heat from the subtropical waters, its thunderclouds becoming more and more organized, forming into tighter spiral bands of densely packed clouds. Joaquin was rumbling to life.

As the storm system turned south, it traveled over waters that were about two degrees Fahrenheit warmer than normal, accord-ing to the NHC. In fact, the water was the warmest on record for those dates in that area. And as this warm water evaporated up into the low-pressure system, it alchemized into billowing, puffy white and gray cumulonimbus clouds. Joaquin was becoming a nearly perfect heat engine. As the water vapor condensed into these clouds, heat was released, which rose to form more clouds. The heat up high lowered the surface pressure of the storm's center. After rising, the warm air was pushed out and away, where it cooled and dropped, then was sucked back into the low pressure at the base of the storm. This rushing air created friction and wind, which began spinning around the central column of clouds. Surrounding air also rushed into the low-pressure center. The direction of the wind is guided by the earth's rotation and the ensuing Coriolis force, which determines which way the wind blows—counterclockwise in the Northern Hemisphere, clockwise in the Southern Hemisphere. Joaquin was recycling its own energy to grow bigger and stronger.

By the night of Tuesday, September 29, Joaquin's winds were reaching 70 miles per hour, nearly double what they had been at five o'clock that morning. The National Hurricane Center was predicting that the storm would become a hurricane by morning. But its forecasters were having trouble anticipating its track.

"The environmental steering currents are complex and are not being handled in a consistent manner by the forecast models," the NHC's 11:00 P.M. discussion post noted. Three of the computer models used by the NHC showed the storm strengthening during the next few days, which meant "the NHC forecast could be somewhat conservative."

Chapter 5
THE OLD BAHAMA CHANNEL

In Florida, TV weather forecasters were talking about Joaquin as an aside, an offshore system that wasn't likely to make landfall. This was one of those storms the meteorologists say they're keeping an eye on, but aren't likely to ruin your weekend plans. Todd Kimberlain, the NHC forecaster who had first tagged Joaquin as Tropical Depression Eleven, had taken a few days off work for a previously scheduled trip to visit friends in New York City and from there travel out to Montauk, on Long Island. Somewhere between the plane and the train, he checked in on work.

"I remember thinking, 'That's not what I forecast,'" Kimberlain recalled, when he noticed that the storm was gaining strength. "It went through this period of intensification that was totally unforeseen." Not only had he not forecast it; neither had any of the forecasters after him.

On Tuesday, as he hit a Long Island beach with friends, the

models still showed the storm heading north, even as it headed south. Kimberlain, ever the weather nerd, has a photo of something he wrote in the sand at the waves' edge: "Joaquin could be coming."

"The models were not handling the track well," Kimberlain explained. "We didn't have a good depiction of the steering flow around the tropical cyclone."

Back in Miami, NHC forecasters were scrambling to figure out what was going on—and how their models had erred. The storm was intensifying so quickly Tuesday evening that the NHC requested a hurricane hunter, an Air Force Reserve plane equipped with weather-recording instruments, to investigate. The planes, depending on the mission, fly over storms, around them, or directly through their eye walls at about ten thousand feet or more. They then scatter dropwindsondes, small devices on tiny parachutes, into the storms to measure wind direction, speed, temperature, height, etcetera.

By Wednesday morning the Air Force confirmed that Joaquin's winds were now blowing 75 miles per hour, the threshold for a Category 1 hurricane, and that the storm was moving southwest at about six miles per hour. Hurricane-force winds extended in a 30-mile radius from the eye. Tropical storm–force winds extended outward in a 125-mile radius.

The NHC issued warnings for the Bahamas. Its 11:00 A.M. discussion post, however, detailed the lack of agreement about the storm's track generated by the different models.

The majority of the models forecast that Joaquin would turn west and curve north within forty-eight to seventy-two hours. These tracks predicted the storm would move inland over the mid-Atlantic states and merge with a trough developing there. However, the European Centre for Medium-Range Weather Forecasts predicted that the storm would stay at sea, moving farther south before curving west and north during the next forty-eight hours.

In light of these differences, the post added a caveat: "Confi-

dence in the details of the track forecast late in the period remains low, since the environmental steering currents are complex
and the model guidance is inconsistent."

Wednesday morning, September 30, dawned calm and bright
aboard the *Minouche* as it sailed south on the Old Bahama Channel. The channel is a deepwater trough between the landmasses
that make up the Bahamas island chain and Cuba. It can be
tricky to navigate in parts, with shallow-water shoals that must
be avoided. But not only is it the most direct route from South
Florida to the Caribbean, it's also protected on all sides by islands that help block winds and big waves.

The *Minouche* was making good time as it motored at a steady
seven to eight knots toward Port-de-Paix. The wind blew westerly at about four knots, giving the sea a slight chop. The auto
forecast along the course the ship would take—past Santaren
South, Dog Rocks, and then Diamond Point—showed clear
skies and no seas. They were right on schedule.

At five o'clock Wednesday morning, the NHC forecast that
came across the *Minouche*'s Inmarsat C terminal said, "Joaquin
is moving toward the west-southwest near 6 mph (9 km/h), and
this motion is expected to continue through tonight. A turn toward the west and a decrease in forward speed are forecast on
Thursday."

Captain Gelera noted the news but remained untroubled by
it. He didn't like that there was a hurricane in the region, of
course, but the storm was holding steady far north of him, above
the central Bahamas. The *Minouche* was now almost two hundred miles southwest of Joaquin, headed in the opposite direction from where the forecasters were predicting it would move.
At this point he was doing what captains do when faced with big
storms: sail away from them.

Gelera knew storms. After joining the merchant marine as a

teenager in his native Philippines he had spent the next four decades traversing the world's oceans. (After China, the Philippines, a much smaller country, is the largest supplier of sailors to the world's commercial shipping fleets. There were 244,000 Chinese sailors in 2015, according to the United Nations' 2016 Review of Maritime Transport, compared to 215,500 Filipinos.)

Gelera came from a big family—he had two brothers and three sisters. His father, a logging supervisor, was killed by a falling branch when Renelo was five years old, leaving his mother, a seamstress, to support herself and her children. "This was hard on my mother, to raise six children without any dad," he said. The family had some land around their house in Tangub City, on one of the country's southern islands, so they grew coconuts, corn, and mangoes to eat and sell.

They were poor, but not in spirit. Gelera's mother guided her children in strict devotion to the Catholic Church. "Every day we go to the church," he said. "Every day." All the boys served as sacristans into their teens, at which point they applied to the seminary together. "Unluckily, due to necessities, my mom told me, 'How are we going to survive if three of you become priests?'" Gelera recalled. "So, I listened to my mom and decided to continue my school."

While his elder brother continued on with his divine studies, Renelo, the middle son, reluctantly resigned from the seminary and went to a merchant marine school in Cebu City, a bustling port town. His younger brother, likewise, went off to study mechanical engineering. (He now lives in Dublin, Ireland.) Even to this day, Gelera remains disappointed that he had to abandon his religious studies. The family's hopes to have a priest among them were crushed when the elder brother died in a motorcycle accident at age twenty-two. But the church remained at the center of Gelera's existence. Throughout his adult life, he would go on Bible retreats whenever he was back in the Philippines. He prays daily, and no matter what ship he is on, he brings a small statue of the Santo Niño aboard and affixes a black-and-silver cross over his bed.

At eighteen, he took his first job on a Philippine passenger ship as a cadet. After less than a year, he became an able seaman on a Greek cargo ship. He made third officer by twenty, second officer by twenty-four, and earned his unlimited ship captain's license at twenty-six. This was a quick ascent. He was very young and his age became an issue. He had rank not only over sailors who had been classmates back at the academy, but men twice his age. He learned the necessary mindset. "When we are outside, I am your friend; when we are on ship I am your captain," he summed up.

Over the course of his career, Gelera has sailed on Greek ships, Japanese ships, Chinese and British ships. He was an officer for Crowley, the American company, sailing routes from Port Everglades, Florida, to Honduras and Guatemala. It was in Guatemala where he met a pretty young lady named Ana Santa Maria. They married in 1988 and together raised three children, who have all gone on to do well themselves. His elder daughter is a medical doctor, his son is a marine biologist, and his younger daughter is a graphic designer. When the children were young, he worked as a harbor pilot in Guatemala so he could stay close to home. As they grew older, he signed on as a captain for cargo ships in Miami. He's sailed the waves of all the world's oceans and many of its seas, including the Mediterranean and Baltic. He has worked hard, and it has given him a good life. By any standard this would be a success, but doubly so for a fatherless boy from an impoverished childhood in the Philippines. He visits his mother every year and sends her money every month.

So the sea provides, but it also takes its toll. Gelera is away for months at a time, and he can't always control when he gets to be home. Holidays and birthdays are hit-and-miss. And of course there's the physical danger of a life on the ocean. During a storm off the coast of Panama, he was caught in heavy waves and high winds. While he was holding the ledge under the window in the bridge for balance, the plate glass vibrated so violently it cracked. A jagged piece fell like a guillotine, lopping off his right thumb.

He learned, as do all merchant marine officers, how to ride out a big storm. He learned how to steer the ship so it faces the swells, doing everything possible to avoid taking a big wave broadside, because you have much more control when the ship pitches than when it rolls. (He has never had a problem with sea-sickness, a significant blessing for a career on the water.) And he learned that soft art of knowing when to gun the engine into the waves and cut the throttle as you ride down the back.

During Hurricane Katrina in late August 2005, he was rounding Cape San Antonio, Cuba, sailing a container ship to Mobile, Alabama, when he found himself about ninety miles behind the storm. The winds came whipping around from the northwest strong enough to push the ship back. So he slowed down and let the storm get farther ahead of him.

He's been on ships in winds powerful enough that they blew containers off the deck. He's navigated unreported icebergs off Newfoundland and rocked and pitched in gales and hurricanes the world over. And on this morning along the Old Bahama Channel, there was nothing in Joaquin's track that made him think he needed to change course. They were sailing away from the storm and were scheduled to reach Port-de-Paix the next day, Thursday, October 1. But a veteran captain like Gelera knows that on the water, complacency kills. There is never a time to take the ocean for granted.

Also in the region, sailing up from the south, was the six-hundred-foot-long cargo ship *Falcon Arrow*, owned by the Norwegian company Gearbulk and flying the Bahamian flag. The *Arrow* had departed the Brazilian port town of Portocel on September 14 and was making its way to Charleston, South Carolina, with a load of timber on deck stacked five meters high.

The ship charted a course that skirted the Caribbean, staying in the Atlantic for the entire journey. Once the ship was north of the Dominican Republic, the plan was to sail northwest to Charleston. The *Falcon Arrow* would have passed *El Faro* going the opposite direction. But on the morning of September 30, the *Falcon Arrow* found its course blocked by Joaquin.

The captain, Ranjit Gokhale, a forty-year sailing veteran from India, had been monitoring the storm since Sunday, as soon as it became a low-pressure system. Once it transformed into a tropical storm, Gokhale knew enough about the Caribbean during hurricane season to stay away. As Joaquin blossomed into a hurricane, Gokhale noticed the erratic path the storm was taking. He plotted some course options that would keep four hundred miles between the storm and his ship. That was not only his preference, it was company policy. "If you cannot do that, you try to keep as much distance as possible," said Gokhale, who has been a captain since 1992.

Gokhale, too, had seen his share of huge storms. In 2000, the captain, who is clean-shaven and speaks in impeccable, clipped English, was sailing a six-hundred-foot ship to the Port of Newcastle, Australia, in the East China Sea. When he was just south of Japan's Okinawa Islands, they ran into heavy seas from Typhoon Saomai, which was still hundreds of miles away. ("Typhoon" is the word for hurricane in the Northwest Pacific region.) Although Gokhale had tried to stay out of the storm's way, Saomai surprised him with a little juke move, stalling in front of Gokhale's ship, which had empty cargo holds on this leg of the journey, making her lighter and higher in the water.

Caught in the winds, he tried to motor the ship through, but even with Gokhale's steady hand on the throttle, the storm pushed the ship backward. Twenty-four-foot swells rocked the boat. "The wind was so strong that you couldn't open the door" on the bridge, Gokhale recalled. "If you did, you'd get sucked out." At one point a big swell lifted the stern completely out of the water, allowing the propeller to spin freely without any resistance. This causes violent vibrations that can damage the propeller shaft, so as soon as this happens, the engine is designed to automatically shut down. "I was a bit scared when the engine cut out," Gokhale admitted. "You have to start it manually."

After a tense couple of minutes, the engineers got the engine roaring again. Now the plan was to reverse direction and sail north, away from the storm. To do that, Gokhale had to turn the ship around—extremely dangerous in high winds and heavy seas. If a big wave were to catch the ship broadside during the turn, the ship could roll, or cargo could shift and give the ship a permanent list. "You have to time it right," the captain said, "and keep full power all the time."

They successfully came about, and now, with the wind at their stern, the ship easily made 14 knots and managed to escape the typhoon zone. They fled north, up into the East China Sea, between the Korean Peninsula and China, where they just waited. "There were quite a few ships stopped there, waiting out the storm," Gokhale recalled. It took two days for Saomai to pass clear of the shipping lanes.

So Ranjit Gokhale knew from experience the importance of caution when confronting vortex winds on the open water.

As he looked at the weather reports on Joaquin, he was surprised to see such wide margins of error for the storm's track. He decided the prudent thing to do was the same thing he'd done during Saomai: delay his arrival and wait. He would either sail a longer and more southerly course along the Old Bahama Channel or, if he had to, duck into the Windward Passage and take cover in the lee of Cuba's eastern tip.

Not too far from the *Falcon Arrow*'s position that day was the 270-foot U.S. Coast Guard cutter *Northland*. ("Cutter" is a general term for Coast Guard ships over sixty-five feet with living quarters on board.)

The thirty-one-year-old ~~Northland~~ ~~carried a crew of 104 and~~
was on routine patrol for drug interdiction in the Caribbean.
That particular morning, the ship was off the north coast of
Haiti, patrolling an area called the North Claw. The *Northland*
had left her home port of Portsmouth, Virginia, in mid-
September and was expected to be at sea for about a month.

The captain was a forty-two-year-old Coast Guard commander named Jason Ryan who had made the law enforcement
arm of the service his specialty. The *Northland* prowled the waters around the Windward Passage, looking for suspicious boats
to board. The ship worked closely with helicopter crews from
Operation Bahamas and Turks and Caicos, investigating boats
the air patrols flagged. At one point on this voyage, a helicopter
alerted the ship to a motorboat off the coast of Puerto Rico that
had what appeared to be bales stacked on the deck. The cutter
launched a small pursuit boat. The motorboat fled and, after a
chase, crash-landed on a Puerto Rican beach. The smugglers es-
caped on foot, but the Coast Guard confiscated what turned out
to be drugs, likely cocaine. (Because the drugs were confiscated
ashore, they were turned over to police there.) But not every day
was consumed with the hunt. Other days were easy cruises on
calm waters. The crew would use these opportunities to run
endless drills: man overboard, small boat launch, and so forth.
On their downtime, the sailors would sometimes fish or swim off
the stern of the ship while a Coastie with an M16 rifle stood
shark watch.

As the patrol wore on, Captain Ryan and his crew began paying attention to the low-pressure center south of Bermuda as it
grew stronger and stronger and crept south toward the Bahamas.
Ryan was mindful of the fact that the *Northland* would be in a
good position to assist in the shipping lanes around the Bahamas, if needed. And with one distress signal or phone call from
higher up the chain of command, the *Northland*'s primary mission would change from law enforcement to humanitarian.

The concept of a Coast Guard did not occur organically in this country. Military services tend to be for the literal minded: The army protects a country's land, the navy protects its waters, and the air force protects its skies. The idea of creating a force responsible for this liminal zone where land meets water came to be as a brand-new nation groped about, feeling out its new responsibilities and creating bureaucracies to fulfill them.

In 1790, a newly formed United States of America, desperate for money to keep its government running, formed the Revenue Marine Service to police harbors in order to collect duties on ships and imports, as well as crack down on pirates and smugglers who threatened trade and reduced the tax base. (Eventually this customs/police agency would become known as the Revenue Cutter Service, hence the term "cutter" for a Coast Guard ship.)

The year before, in 1789, Congress had created the U.S. Lighthouse Establishment and put it under control of the Department of the Treasury. The job of what would eventually be called the Lighthouse Service was to bring all twelve lighthouses scattered along the country's eastern, and at that time only, coastline under federal control in order to make sure they functioned without interruption. The primary motivation was humanitarian; the United States didn't want ships crashing off its beaches. But it was also a nod to how vital shipping and trade was to the young republic. The country couldn't afford to have its trade compromised as it tried to earn its place in the world. The growing control of the shoreline also gave rise to the Steamship Inspection Service in 1838, followed by a modest shore-rescue operation ten years later called the Life-Saving Service.

In 1848, the Life-Saving Service assumed control of a network of locally run beachside stations manned by volunteers who helped rescue foundering ships. The new service equipped these stations with rowboats that could be launched through the surf, small cannons called carronades that could shoot a ball with line attached out to a ship (so sailors could pull themselves to shore), and signal rockets to warn ships that they were getting too close to dangerous waters. But they still relied on volunteers. It took a Category 3 hurricane that struck the Georgia–South Carolina coast in 1854—drowning twenty-six people—for the government to fund a full-time keeper for each station.

By 1915, this hodgepodge of services was all rolled into one— the United States Coast Guard. Its official motto may have been *Semper Paratus,* Latin for "Always Ready," but its unofficial motto was "You have to go out, but you don't have to come back." This was a grimly laconic reading of an old Life-Saving Service regulation that stated a station keeper "will not desist from his efforts until by actual trial, the impossibility of effecting a rescue is demonstrated. The statement of the keeper that he did not try to use the boat because the sea or surf was too heavy will not be accepted, unless attempts to launch it were actually made and failed." In other words, a guardsman was expected to do everything, including risking his own life, to attempt a rescue. The service no longer officially encourages that kind of thinking— missions will not unduly risk a crew's life. But privately, Coast Guard search and rescue specialists will tell you they'll push the limits as far as they can.

Today the Coast Guard is the only branch of the military that, during peacetime, is exempt from the Posse Comitatus Act, which prohibits the armed forces from taking part in searches, seizures, and arrests of U.S. citizens. By federal law, the Coast Guard is both a military service and a law enforcement agency. Its duties include policing our coastal waters (which extend to about two hundred miles offshore) to enforce maritime laws, like fishing regulations, and far beyond those boundaries to interdict perceived threats such as drug smuggling. The Coast Guard also oversees inspections of all vessels operating in U.S. waters for seaworthiness. It enforces safety regulations, as well as domestic laws and international treaties regarding marine conservation. It responds to marine emergencies and investigates the causes of accidents. The disparate roles of its predecessors have spread the Coast Guard thin, forcing it to straddle many duties over countless nautical miles.

But at its core, the Coast Guard is about saving people. So as Captain Ryan looked at the storm hovering over the busy waterway, he prepared his ship and crew for what was increasingly looking like an abrupt change in mission.

Joaquin seemed to be gaining strength by the minute. The NHC's 5:00 P.M. report on Wednesday, September 30, estimated 85-mile-per-hour winds. By 11:00 P.M. the hurricane's winds had reached 115 miles per hour. That's a jump from a Category 1 to a Category 3 hurricane in just six hours.

Hurricanes that size are several miles high, reaching through the troposphere and poking into the stratosphere, where the condensation freezes into cascading sheets of ice crystals. The mushrooming cloud cover will typically spread more than one hundred miles across the sky, and contain an eye of calm in the center that is anywhere from five to thirty miles wide. The wind can churn up a huge wall of water near this eye, and if the storm makes landfall, this wave can inundate entire cities.

Yet birds caught in the eye can survive a hurricane, released from its grip hundreds of miles from where they started.

A hurricane is one of nature's most powerful forces. A Category 1 storm blowing 90 miles per hour over a forty-mile radius could power a quarter of the world's electrical grid for as long as the storm was blowing. An average-sized hurricane drops as much freshwater as the entire human population uses in a day.

But there is no need for abstractions and hypotheticals to understand the power of a hurricane. History is littered with real-life examples.

The word "hurricane" comes from the Mayan and Taino Indians of Central America and the Caribbean, whose myths referred to a god of chaos and wind named Huraca'n, or Juracán. Drawings and carvings depicting the deity feature a head with two arms sweeping in a counterclockwise direction. This closely resembles the eye and wind direction of a hurricane. MIT professor and hurricane expert Kerry Emanuel notes in his book *Divine Wind: The History and Science of Hurricanes* that this is curious because hurricanes are far too big for humans to perceive from the ground. It was only when planes and satellites took us above the clouds that we were able to get an overhead view of what hurricanes look like. So how did these early Americans come up with this image? Perhaps they divined the shape and movement of a hurricane from the pattern of destruction left behind, Emanuel speculates. Or maybe they based their depiction on water spouts, smaller tornado-like funnels that mimic the larger storm's physics.

It wasn't until the 1800s that Westerners discerned that hurricanes were wind vortexes. In 1821, an amateur meteorologist named William Redfield examined the aftermath of a hurricane in Connecticut. He noted that on the storm's eastern side, trees had fallen to the northwest, while on the storm's western side the trees had fallen to the southeast, indicating the storm had a circular pattern of destruction.

"Hurricane," by the way, is simply a regional name, the one used in the Atlantic and Northeast Pacific. In the Northwest Pacific, as previously mentioned, they are called typhoons. In the South Pacific and Indian Ocean, they are known as cyclones. These are all the same phenomenon: a powerful wind vortex. And all fall under the general term "tropical cyclones."

By whatever name, vortex winds have thwarted the ambitions of empires throughout history.

Twice in the thirteenth century, Mongolian emperor Kublai Khan tried to invade Japan from the Korean Peninsula. Twice his armadas were destroyed or turned back by typhoons. Japan's emperor called the storms Kamikaze, or "Divine Wind."

Three centuries later, as Europe's powers battled to control the New World, France and Spain were in a contest to claim Florida for their respective monarchs. Both had colonies on a peninsula that neither wanted to share. In 1565, a fleet of eleven Spanish ships sailed to attack France's Fort Caroline on the St. Johns River near what is now Jacksonville. But a hurricane bore down on them, and only five ships made it back to port. The French tried to take advantage of the weakened Spaniards, setting sail with a force of about four hundred men. They, too, were met by a hurricane, scattering their ships. The Spanish counterattacked by land and overran the French, giving Spain control of Florida.

This would happen again and again as the European powers divvied up the New World, long before we started giving these storms names. In 1640, a hurricane ripped apart a Dutch fleet sailing to attack Havana, Cuba. In 1666, a British fleet of seventeen ships was destroyed by a hurricane in the Lesser Antilles. These two events helped cement Spanish control of Cuba and French control of Guadeloupe.

Three back-to-back hurricanes in October 1780 killed tens of thousands of people and sank dozens of ships throughout the Caribbean and United States. The storms severely weakened the British Navy as it fought the American revolutionaries. The first storm struck Jamaica and then tore through Cuba, sinking British warships and collapsing whole towns, killing an estimated 3,000 people, half of them sailors. The second storm sped from Barbados to Bermuda, claiming roughly 4,300 lives. On the island of St. Vincent, a twenty-foot storm surge washed villages out to sea. On St. Lucia, the storm killed 6,000 people. About 9,000 more died in Martinique. Fifteen Dutch ships sank off Grenada. All told, this was the deadliest storm on record in the Western Hemisphere. Incredibly, another storm picked up within days and ripped through the region, smashing sixty-four Spanish warships sailing to take back the Florida panhandle from the British. Nearly 2,000 men died.

By 1898, as America prepared to fight Spain for control of Cuba, the threat hurricanes posed to naval armadas was thoroughly established. Willis Luther Moore, the chief of the newly minted U.S. Weather Bureau, recounted a meeting that he and his boss, Secretary of Agriculture James Wilson, had with President William McKinley to lobby for the creation of outposts throughout the Caribbean to provide early warning of hurricanes.

"I can see him now as he stood with one leg carelessly thrown across his desk, chin in hand and elbow on knee, studying the map that I had spread before him," Moore recalled, speaking of McKinley. "Suddenly he turned to the Secretary and said: 'Wilson, I am more afraid of a West Indian hurricane than I am of the entire Spanish Navy.' To me he said: 'Get this service inaugurated at the earliest possible moment.'"

This became the predecessor of the National Hurricane Center.

The incipient service had some growing pains with tragic consequences, namely Galveston, Texas. In early September 1900, Cuban forecasters warned their American counterparts that a massive hurricane had just barreled past Hispaniola, Jamaica, and Cuba, and was now on its way to the Gulf of Mexico. As documented in Erik Larson's 1999 book *Isaac's Storm,* the Americans ignored the Cubans, who they may have viewed as inferior and excitable. At one point, irritated U.S. weathermen blocked the cables from Cuba.

Based on a flawed analysis, American forecasters predicted that the storm would curve north up the East Coast of the United States. As the Weather Bureau warned boaters and fishermen in New Jersey to stay in port, the hurricane made landfall in Galveston on September 8, 1900. The death toll was an estimated ten thousand people.

Chapter 6

"WE'RE GOING INTO THE STORM"

It's one of the most fundamental and obvious rules of the sailor's life: Monitor the forecasts, heed the NHC's advisories, and if there's a big storm in your path, change course. Simple.

But as the *American Practical Navigator*—a mariner bible written by Nathaniel Bowditch, first published in 1802 and continually updated ever since—points out, these bulletins and forecasts are not infallible. Sometimes they "may be sufficiently in error to induce a mariner in a critical position to alter course so as to unwittingly increase the danger to his vessel."

In a case like that, sailors need to know how to spot the early signs of a hurricane regardless of what the forecasts say. The first thing to be on the lookout for is an exceptionally long swell, which in deep water comes from the direction of the storm and can ride along days ahead of it. Then cirrus clouds will appear overhead. Where these clouds converge in the distance is a good indication of the storm's location. After the cirrus clouds drift into sight, the barometer will start a long, slow fall. In technical language so precise it verges on poetry, the *American Practical Navigator* states, "The cirrus becomes more confused and tangled, and then gradually gives way to a continuous veil of cirrostratus."

With the advantage of early detection, a shipmaster should be able to steer clear of an approaching storm, either by sailing around it or simply outrunning it. If the ship's position makes it impossible to avoid a hurricane, the master can plot to steer the ship to the safest part of the vortex, the "navigable semicircle." This is the section to the left of the storm (as viewed from above) in the Northern Hemisphere. The section to the right is known as the "dangerous semicircle," because that is where the full cyclonic force of the rotating wind comes from. The navigable semicircle sits on the less powerful arm of the rotation. If you're a shipmaster who has no choice but to confront a hurricane, this is where you should steer.

EL FARO, WEDNESDAY, SEPTEMBER 30, 2015
4:00 A.M. to 8:00 A.M. Watch

Even as Captains Gelera and Gokhale monitored the storm and made plans to stay away from it, Captain Davidson was charting *El Faro*'s course straight for Joaquin.

It's not that Davidson and his officers were unaware of the available information. On *El Faro*'s bridge at 6:00 A.M. Wednesday, the captain and his chief mate, Steven Shultz, were already focused on Joaquin, looking at weather reports and discussing their options. (A voice and data recorder picked up all conversations from six microphones placed around the bridge.) The ship had traveled about two hundred miles overnight, putting it in the Atlantic about one hundred miles northwest of the westernmost Bahamas. Captain Davidson clearly thought he could skirt south of the storm's center and still be safe.

The bridge was continuously staffed by one of the ship's officers and an able seaman (AB) who served as helmsman in a cycle of three four-hour watches. The officer's job was to carry out the captain's navigational commands, and the able seaman steered the ship as directed. The first watch, 4:00 A.M. to 8:00 A.M., was manned by chief mate Shultz and AB Frank Hamm. Third mate Jeremie Riehm and AB Jack Jackson would handle the 8:00 A.M. to noon watch, followed by second mate Danielle Randolph and AB Larry Davis from noon to 4:00 P.M. At that point, the cycle started over again. The officers and seamen maintained the same shift sequence throughout the journey, whether it was morning, afternoon, or night. The captain did not stand a prescribed watch. In a sense, he was always on duty.

Davidson had a very capable chief mate, and it was clear the captain respected him. Shultz was a lifelong mariner, a 1984 graduate of the U.S. Merchant Marine Academy in Kings Point, New York, and a Navy veteran who still served in the Navy Reserve. Shipping was his life, and had given him the love of his life, too. In 1993, Shultz was crewing a ship docked in Santos, São Paulo, in Brazil, when a young woman called asking permission to come aboard. She was looking for her father, a stevedore unloading the vessel. Shultz, with warm brown eyes and a mustache, came down to get her.

"He was very sure of himself," Claudia Shultz recalled. "He came down in his work clothes and said, 'I'll go find your dad.' We talked a little bit. I told him my dad and uncle were on the ship, and if they saw me talking to him, they'd kill him." He laughed. All the men in Claudia's family were in shipping, either sailors or stevedores.

After one date, "We decided it was love at first sight," she said. That was in November 1993. They were married two years later. By the time Shultz set sail on *El Faro* in late September 2015, their daughter was in high school and their son had graduated college and was about to start working on his master's degree in English literature. Shultz had never stopped sailing for a living, first on international routes, then domestic ones with TOTE, which kept him at sea for ten weeks at a stretch instead of four months. By Claudia's account, he still loved his job and the life it afforded him. Until this trip. Normally he sailed *El Yunque,* but for some reason he had been reassigned. This was his first run on *El Faro.*

"He was off," Claudia recalled. "He just did not want to make that trip."

In bed the night before he left, he couldn't sleep. "Don't you want to get some rest?" Claudia asked him. "He said 'No, I just want to lay here and watch you sleep,'" she remembered him saying. *I don't really want to go,* he told her. *I want to stay with you and the kids.* But he knew he had a job to do, and in the morning, he left.

"I'm anxious to see the new BVS," Shultz told Davidson on the bridge at 6:03 A.M., referring to Version 7.0 of the Bon Voyage System, a computer program and weather forecasting and reporting service TOTE used. Davidson was also a fan of the BVS, which enhances NHC forecasts with other weather data such as fronts, tides, currents, and significant wave heights. But these enhancements took time. The National Hurricane Center updates its forecasts every six hours, with supplemental reports every three hours during a hurricane watch, and the BVS additions took a few hours more.

Davidson, however, was unaware of this delay and appeared to think the BVS forecasts were as current as the NHC's. That may be because the BVS was advertised as providing "real-time data." Moreover, the BVS system was available only on the computer in the captain's office—normally a minor inconvenience, but one that would prove significant on this journey. To get the BVS data on the bridge, the captain had to forward emails from his computer. (The bridge did have an Inmarsat C terminal programmed to print out NHC priority weather messages, a seafaring requirement.)

"I'm connecting right now," Davidson said as he signed on to his email account on the bridge's laptop computer, set up on the chart table behind the ship's main console and the helm. From the inside, *El Faro*'s bridge resembled a control room in a factory more than the nerve center of a ship. Industrial beige walls surrounded a gray metal console containing computer screens, switches, and indicator lights, set against a series of windows looking out over the main deck and the bow. A wheel was set in the middle of the console about waist high. A metal rail ran the length of the console so that sailors could hang on in rough weather. Behind the console was the chart room, which could be separated by a curtain and contained a long metal table where maps could be laid out. It had drawers that held the charts and places to hold pencils, compasses, and dividers for charting on the maps.

As Davidson logged on to the bridge's laptop that morning to retrieve the BVS email package he'd forwarded, he had no way of knowing he was getting old information—and not just due to the BVS's customary lag time. In this case, the BVS had accidentally repeated the NHC's forecast information from six hours earlier.

The two men hunched over the computer screen were reviewing information that was completely out of sync with where they were now and where they were going.

"This is forecasted to go north," Davidson said.

"The worst weather's up here," Shultz noted.

Outside it was still dark, the rising sun just starting to lighten the sky, the seas calm. The bad weather, which they believed to be about five hundred miles to the north and east of them at that moment, must have seemed an abstraction. In reality, Joaquin was closer than they thought, a little farther west and also far-

ther south, and moving south, not north—on track to intersect
the ship if it remained on its southerly course. The storm was
hoovering up all the clouds around it, packing them into the
vast, dense cloud cover that was mushrooming out over the center.

"So we'll just have to tough this one out," Davidson declared
with finality. His thinking seemed to be set: There would be no
significant deviation from the course. He expected some rough
seas, but as he would repeatedly tell the crew, nothing he hadn't
seen during his years in Alaska.

Shultz, examining the charts, mentioned the option of cutting south through passages between the islands to the Old Bahama Channel. But he didn't think they needed to make a
decision just yet. "I would wait, get more information," the chief
mate said.

"This is why you watch the weather all the time, all the time,"
Davidson said at 6:15 A.M. He cleared his throat. "Absolutely."

The captain and his chief mate stayed in the chart room to
review their options. At the big metal table, they hunched over a
large nautical chart of the area. They discussed Joaquin, which at
that point was a tropical storm, went over the weather reports,
and scoured the map as they plotted alternate routes they could
take if necessary. While Davidson remained committed to staying the course for the time being, he was open to making slight
adjustments to avoid the worst of the weather.

Captain Davidson peered at the map and pointed along the
projected track line of Joaquin. "We're south and west of it," he
said, noting a future position of the ship.

The two men studied the storm's track in relation to their
itinerary and considered a new course that would take them farther south but still keep the islands on their starboard side—that
is, stay in the open Atlantic, as opposed to ducking into the Old
Bahama Channel. "Yeah, you can steer a little bit more away
from it," Davidson said. "A little more away from the center."
They entered longitude and latitude coordinates to create track

lines from one waypoint to another. "Connect those two dots. . . . Neaten this up." . . . "Yes sir." They gave the waypoints names, alpha and bravo, and then talked about running the route through a Mercator projection, which allowed sailors to plot a course on a curved earth and see it as a straight line.

"What's helping us right now is our speed," Davidson said, believing that they would motor past the storm quickly, limiting their exposure to the high winds and waves. *El Faro*'s maximum speed was 24 knots, which was a decent clip for an older ship.

Joaquin was traveling south at only six miles an hour. But the vortex winds spinning around the eye were growing stronger. At the time the two men were going over their charts, the storm's winds were reaching 70 miles per hour and Joaquin was undergoing its period of rapid intensification as it moved over 86-degree water, the warmest on record for that time of year.

Ultimately, the men decided to sail sixty miles south of where they thought the storm's center would be when they crossed its path. In reality, because the captain was referring to an hours-old weather forecast, the storm would already be south of them at that position, not north, and forty miles closer.

As the other captains in the area sought two-hundred- to four-hundred-mile buffer zones between their position and the weather system, it's remarkable to hear Davidson plan what he thought would be a sixty-mile jog south of Joaquin. Even without accounting for the forecast errors they were using in their calculations, sailing that close to a strengthening tropical storm was extremely risky. To sailors such as AB Kurt Bruer, who had crewed with Davidson previously on *El Faro*, it showed a lack of respect for the Caribbean, which can be soporifically calm for months and then erupt into violence.

"He was not experienced handling storms in the Caribbean," Bruer said. "Alaska and the Caribbean are two different things." In the Caribbean, he said, "you only get a little room for error. Once you pass that point, there's nowhere to turn. You have two options, turn the ship around, which takes a lot of time and fuel, and the company's not going to be happy with, or continue. He made a choice and he took a gamble. Most captains I sailed with in the Caribbean, as a rule of thumb, stay three hundred miles from the storm."

As captain, Davidson had full authority to make this decision. TOTE's policy was to give all operational control to the ship-masters once they left port. The only thing the company expected was for the ships to stay on schedule. Unlike other shipping companies, TOTE had no specific protocol for monitoring the positions of its ships once they were under way.

"What we'll do is, once we know [that] this storm is clearin' up outta' here, we'll be more assertive towards getting back to the optimum track line," Davidson said.

"Yeah," Shultz agreed.

"I think that's a good little plan, Chief Mate," the captain said at 6:28 A.M. "At least I think we got a little distance from the center." Then, a moment later, "Nuts on, as we say."

At that point, the captain announced he was headed to the engine room to tell the chief engineer to expect heavy weather. He popped back into the bridge ten minutes later, explaining that he'd stopped in to tell the crew in the galley to secure dishes and condiments. "You don't want to wake up with maple syrup and cayenne pepper mixed together on the deck," he said.

After half an hour, Davidson announced he was leaving the bridge again, saying, "Let me go down and check up on everybody."

8:00 A.M. to Noon Watch

At 7:44 A.M., third mate Jeremie Riehm arrived on the bridge to get briefed before the start of his four-hour watch with helmsman Jack Jackson.

"You hear about the storm?" Shultz immediately asked Riehm.

"Yes, yes, I'm aware of it, caught a little bit on the news," Riehm said. "It's gonna be a hurricane tomorrow or later today according to the Weather Channel."

Shultz showed Riehm the route number logged into the GPS, route number 14, and noted the waypoints A to B, indicating a position about ten miles off Elbow Cay and San Salvador Island in the Bahamas. Then Shultz showed Riehm the storm's overnight progression.

"It was up here last night, but now it's down here," Shultz pointed out, referencing the flawed BVS forecast. Then, talking about their original route's projected intersection with the storm, Shultz said, "So that was a direct hit. I mean—our timing was perfect to the eye." Here Shultz pointed out the alternate route he and the captain had plotted earlier. "Now, worse comes to worse, we can duck in through one of the passages between the islands to the Old Bahama Channel."

"Yeah," Riehm said.

"Secure for sea. Heavy weather," Shultz intoned before leaving the bridge. "Get your legs on."

If anyone's sea legs were strong, it was Riehm's. He had been raised for much of his youth on a sailboat in Miami, by free-spirit parents who had little use for conventional thinking. Eventually the family settled on terra firma, in a home on Pine Island, off Florida's west coast, where Jeremie and his brother continued to sail and boat. At one point Jeremie sailed alone from Florida to Mexico. After a brief flirtation with college in Daytona Beach, he quickly realized a world spent indoors was not for him and decided to go to mariners school.

But before he left Daytona he met Tina, a vivacious coed from a tough part of Camden, New Jersey, raised by a single mom on welfare. Riehm had been dating one of her friends. The three of them often hung out together, and one day the friend confided in Tina that she wanted to break up with Jeremie. He was too nice, she said. Tina thought this was insane. Not only was he nice, he was handsome and funny, too. After the breakup, Tina swooped in and the two had a passionate romance that gave them their first child, out of wedlock. Jeremie, in love, proposed. Tina didn't say no, but she didn't say yes, either. Having grown up poor, she was determined to finish her education and worried that family life would take her off that path. She told Jeremie, *Not until after I graduate*. Every year for the next four years he proposed, even getting their baby daughter, Clancy, in on the act when she was four. Finally, in the fifth year, degree in hand, she accepted.

The couple was a study in contrasts. He was white, she was black. He grew up on the water. She grew up in the inner city. For two people curious about the world, they had a lot to learn from each other. That made the union work.

Over the course of their four-year courtship, Jeremie enrolled at the Harry Lundeberg School of Seamanship, a school for un-licensed mariners, in Piney Point, Maryland. After Tina gradu-ated, Jeremie moved her and Clancy to Pine Island, to be near his family. Soon after, the couple had another baby, a boy named Carlo. Meanwhile, Jeremie began shipping out and building his career. Riehm was rare among *El Faro*'s officers in that he had not attended a merchant marine academy, whose students auto-matically become officers upon graduation. Instead, motivated by his desire to provide for his young family, he studied on his own for his officer's license.

So he knew the water, and he knew sailing, and, alone with Jackson on *El Faro*'s bridge, Riehm knew exactly what they were in for.

"We're gonna get slammed tonight," Riehm mused at 8:15 A.M.

"Yeah," Jackson replied. "I was thinking that all through Au-gust, like it's been too quiet this season."

At about 8:30 A.M., the captain returned to the bridge. "So we got a little weather coming in, I'm sure you heard," Davidson told the two sailors. "Tough to plan when you don't know, but we made a little diversion here. We're going to be further south of the eye. We'll be about sixty miles south of the eye." Then, as if he was trying to convince himself, Davidson added: "It should be fine. We are gonna be fine. Not 'should be.' We *are* gonna be fine."

Neither sailor said anything to that.

Outside the seas were mostly calm, with a moderate chop building. The sun shone strong amid clear skies. Riehm noted that the barometer was rising. Usually when hurricanes are hundreds of miles away, as Joaquin was at that point, the barometer rises slightly and the sky is clear of clouds.

By this point, Joaquin was a hurricane. The Air Force hurricane hunter had flown over the storm and logged wind speeds of 75 miles per hour near the eye. NHC forecasters were expecting the whole system to continue strengthening over the next forty-eight hours.

The sailors on *El Faro*'s bridge didn't indicate that they had seen this latest update—not that it would have changed anything. Davidson was clear-eyed about what he was doing. "We're going into the storm," he said at 9:22 A.M.

Davidson's confidence in his ship was rooted in specific and recent experience. He had taken it through Erika, a tropical storm that had blown across the Dominican Republic only a month earlier, in August and had prompted him to take the Old Bahama Channel. "That's the first real storm I've been on with this ship," Davidson told Riehm and Jackson. "Ship's solid."

Riehm agreed—to a point. "This ship is solid. It's just all the associated bits and pieces," he said. "The hull itself is fine. The [power] plant, no problem." It was all the smaller stuff that can break free in high winds that he was worried about.

"We'll see," Davidson mused. Maybe the impending storm shook loose the memories, or maybe he was trying to build his team's confidence by letting them know he had been through worse and they were in expert hands. In any event, he started talking about his Alaskan experiences. "The worst storm I was ever in was crossing the Arctic," he said. "Horrendous seas." The waves were so bad, he said, that the ship would ride up, slide down into the trough, and then roll. "You go up and you'd come down and you'd crash, and all your cargo would break loose." Cars were sliding around, he said, and sediment in the ship's fuel tank got churned up and clogged the lines.

"It was like that for a solid twenty-four to thirty-six hours," Davidson recalled. "It was bad. We had a gust of wind registered at one hundred and two knots. . . . It was the roughest storm I had ever been in—ever."

Riehm, who had seen a storm or two as well, likely nodded in agreement with the captain. Then he asked, "I know I should know this, but when's our next BVS update?"

Davidson told him to check the computer. "Every six hours," Riehm said, answering his own question. "Another one at ten."

The Polish crew was working by now, and the captain could barely hear his sailors above the racket outside. The welders' power tools whined on the decks, grinding metal as they refitted the ship's winches with new cable trays.

"Yeah, I'm worried Team Poland is not ready for this—the weather," Riehm observed. "They got big pipes just propped up, laying against the bulkhead."

The Poles, he said, "just kind of do their welding and they kind of do their yard work and they expect the cleanup and all the securing . . . just to happen for them."

"Yeah," Jackson said. "Oxygen bottles just rolling around everywhere."

"Team Poland," Jackson snickered. They both laughed.

Like Riehm, Jack Jackson had also trained at the Harry Lundeberg school in Piney Point, following high school in New Orleans. A clean-shaven sixty-year-old, with a head of thick rusty-blond hair and an impish smile, he had been at sea ever since. Jackson worked tugboats, transocean freighters, U.S. Navy warships. He had sailed nearly all the seas and oceans of the world, including on grain ships to Africa and a resupply mission to the Antarctic's McMurdo Station. His younger brother, Glen Jackson, recalled Jack coming home on shore leave with the most amazing stories: collisions with whales in the English Channel, watching a sky inexplicably turn fluorescent green during a night watch in the middle of an empty ocean. These stories had impact. When he was finishing high school, Glen told his older brother he wanted to join the merchant marine, too.

"I asked him for help," Glen recalled. "He absolutely forbade it. He said, 'You don't have the temperament.'" It wasn't an insult. Jack was trying to save his younger and more restless brother from a life that would not have suited him. "It was hard for me to appreciate the monotony of life at sea," Glen Jackson said, and what it took to survive it.

To drive home the point, the older Jackson paraphrased Samuel Johnson's quote that going to sea is like going to prison, with the chance you might drown.

Glen heeded the advice and became a wildlife conservation officer instead.

At 10:35 A.M., the Inmarsat C terminal in the bridge clattered to life and printed out the latest NHC forecast advisory. Riehm walked over to the printer and ripped the page off along its perforation.

"It's moving away fast?" Jackson asked.

"Uh, no, it's not moving away," Riehm answered. "We're on a collision course with it nearly."

Noon to 4:00 P.M. Watch

At 11:45 A.M., second mate Danielle Randolph came onto the bridge along with her helmsman Brookie Larry Davis, a sixty-three-year-old former fisherman from Jacksonville. Riehm briefed her on the route change. By now, Captain Davidson had gone belowdecks.

"We are diverging," Riehm told her.

"Oh, yeah, I see," she said, scrutinizing the charts in front of her. Then she cleared her throat nervously. "All right."

"So he wants to get on this track line," Riehm said, showing her the coordinates on the GPS.

"Ahh, okay," Randolph responded, digesting the change in plans.

The two talked some more about the new route and the track lines. Then Randolph, out of the blue, aired the first of what would be many reservations about her captain's decisions.

"He's telling everybody down there, 'Oh, it's not a bad storm. It's not so bad. . . . It's not even that windy out. . . . Seen worse,'" Randolph said.

Randolph had sailed with Davidson before, and she had her opinions about his management style.

Oh, I've got to ship out with him? Randolph had groaned to her mother, Laurie Bobillot, when she learned Davidson would be her captain on this contract. "She did not care for that man," Bobillot recalled. "A lot of times, she would say to me that he was in his stateroom quite a bit, more so than she felt [he] should have been." Randolph said she would have liked to have seen Davidson on the deck more, talking to the crew, trying to understand the ship from topside, rather than just from the electronic controls and data printouts.

Bobillot, who raised her kid to persevere, remembers telling her daughter, "Everybody's got their faults. Kind of suck it up and accept it for what it is."

Yes, I know, Mom, Randolph replied obediently

Randolph wasn't alone in her assessment. While colleagues often described Davidson as meticulous, he was also seen as aloof. In the three times he sailed with Davidson, second mate Charles Baird said he never saw the captain walk the deck to mingle with the crew. According to Baird, Davidson's routine was to visit the bridge first thing in the morning to make sure the ship was on course, the speed was appropriate, and to hear how the previous shift went. Then he'd have breakfast and return to his quarters, where his office was located. "He would visit the bridge at various times of the day to check on things," Baird said. Most other captains Baird worked for would walk the decks frequently, making sure the chief mate had double-checked that the trailers and vehicles were chained down properly, and talking to the bosun, who is in charge of the deck crew, about cargo lashings. "They just wanted to be more involved," Baird said.

Kurt Bruer, the able seaman who had sailed on *El Faro* in February, corroborated the captain's habits.

"He wasn't really walking the ship," Bruer said. "He didn't really take the time to see what the problems were." Other captains, he added, made a point to get to know the crew a little.

(It should be noted that Davidson fired Bruer after their last trip together. Bruer explained that during watch, while in port, he had asked a security guard to cover him while he ran to get a cup of coffee. When he returned, the third mate was there. The mate criticized him for leaving his post, and harsh words were exchanged. The mate complained to the captain, who fired Bruer for insubordination. Bruer has continued to regularly work on ships since the episode.)

As Randolph critiqued Captain Davidson, Riehm stayed focused on the charting, speaking numbers out loud: "One three eight," meaning a heading south at 138 degrees. The crew was aware of the six microphones on the bridge (which are required to record twelve hours of the most recent data, including audio, but can record more than twice that amount, depending on the available data storage), but they also knew those recordings would only be heard if there was a reason to listen to them—that is, a major incident or accident. At any rate, Randolph didn't seem concerned about being picked up by the microphones. She had concerns about Davidson and she was not reluctant to share them.

She continued quoting the captain: "'It's nothing—it's nothing!'" But then, "If it's nothing why . . . are we going on a different track line? Think he's just trying to play it down because he realizes we shouldn't have come this way." She believed the captain was just "saving face," she said.

"We're getting sea swells now," Davis, the helmsman, said.

"Well," Randolph noted at 11:50 A.M., "the storm is here."

Joaquin's wind-pushed waves were in fact beginning to reach *El Faro*. At this point, hurricane-force winds of 80 miles per hour extended about 35 miles out from the eye of Joaquin, with tropical storm–force winds stretching out to a 125-mile radius.

A couple hours later, around 2:00 P.M., Captain Davidson was back on the bridge with Randolph. The storm was now about 120 miles north of the Turks and Caicos and the Bahamas and roughly 300 miles east of *El Faro*, but heading south, blowing a bit harder, 85 miles per hour at the eye. If its current movement—west-by-southwest at six miles per hour—continued, the storm was on a direct path to intercept *El Faro*. But the forecasts still predicted it would move west and then turn north.

What neither the captain nor the crew knew at that point was that the NHC was continuing to have a difficult time tracking the storm. Joaquin was defying the models. In retrospect, the three-day track forecasts were off by 536 miles. The NHC's branch chief would later call this a "one in one hundred type track error." The same was true for the storm's intensity errors. The three-day intensity forecast, meaning the cumulative wind increase over those three days, was 80 knots (92 miles per hour) too low. On that day alone, in fact, Wednesday, September 30, the NHC would be off by a whopping 30 knots (34 miles per hour), the difference between a tropical storm and a Category 2 hurricane.

Part of the problem was that the wind shear, the speed and direction difference between winds in the upper and lower atmosphere, was not behaving as expected. Strong wind shear would cause the thunderstorms to be ripped off the top, weakening the hurricane. Low shear would mean the storm would remain intact. The forecasters saw all the evidence of high shear, but Joaquin wasn't responding to it. The storm seemed resistant to the effects of the shear.

On board *El Faro*, these forecasting errors were compounded by the captain's reliance on the outdated BVS forecasts.

"Do you know if *El Yunque* took the Old Bahama?" Randolph asked Davidson, referring to *El Faro*'s sister ship, which at that moment was returning from Puerto Rico en route to Jacksonville. *El Yunque*, a year younger, came from the same shipyard, and was the same size as *El Faro*.

"She did not," Davidson answered. "They put some turns on her and are staying ahead of her [Joaquin]." He added that *El Yunque*'s captain had seen some gusts to 100 knots.

Randolph used the reference to make light of the fact that *El Faro*'s anemometer, which measured wind speed, was broken. As a result, she said, they'd just have to stick her helmsman, Larry Davis, outside during the blow and see how he held up. "If you get blown off the bridge we'll be like 'Oowww, it was about a hundred, [or] ninety-mile-per-hour gust.'"

Standing at the helm, Davis undoubtedly grinned. The two had sailed together before and they made an interesting pair, the fiery academy graduate and the grandfatherly veteran mariner. But they got along well. He called her "my coffee buddy," because they both took their joe very seriously.

Davidson, meanwhile, assured everyone that they were going to be "far enough south" to avoid the storm. "Going to get a little rougher," he said. "These ships can take it."

"Yeah, they're built for Alaska," agreed Randolph. Despite what she'd been saying earlier about her captain, she maintained a collegial attitude in his presence.

Davidson remained determined not to change course. He had confidence that the ship would slip under the storm as it veered west. Looking ahead, however, he thought Joaquin's residual effects would likely generate big swells in the open water of the Atlantic, so he talked about taking the Old Bahama Channel back from Puerto Rico. But first, he said, he wanted to check with management. Randolph sounded surprised.

"It used to be just, 'We're doing it,'" she said. "'You people are sitting in your office behind a desk and we're out here—we're doing it.' Yeah."

"Well, I'm extending that professional courtesy," Davidson replied, "because it does add 160 nautical miles to the distance."

"Yeah, it also saves on the ship's stress," Randolph retorted.

In the background, the radio blared a weather warning: "The National Hurricane Center has issued a hurricane warning for the central Bahamas. . . . The Coast Guard requests all mariners use extreme caution. . . . The United States Coast Guard aircraft standing by on channel sixteen."

"'Aircraft,' he said?" Randolph sounded alarmed. "Oh, wow."

"Wow," the captain echoed.

Then, after reviewing the weather maps again, Davidson told Randolph, "So, by two o'clock [A.M.] on your watch you should be south of this monster."

Randolph concurred. "Yeah, it seems like we're gonna get through it pretty quick. By tomorrow afternoon."

"Well, I'm gonna go watch a little television, I think," Davidson said.

"Weather Channel?" Randolph asked.

The captain didn't reply. At 2:19 P.M., Davidson exited the bridge.

One of the more curious observations AB Kurt Bruer made about the captain might have explained his reclusiveness. "He liked to play video games," Bruer recalled. "He would always talk about playing *Call of Duty*. I don't know if he did it only on his off-time or what. But in the bridge he'd tell us, 'Yeah, just got done playing *Call of Duty.*' He was really into it." A Coast Guard "shiprider" who had once inspected *El Faro* while the ship was in transit corroborated that Davidson "liked to play video games" when he wasn't on the bridge.

An hour after the captain had left the bridge, *El Yunque* popped up on the radar, thirty miles away.

"They're abeam of us right now," Randolph said. "They're doing twenty-two knots. . . . They're trying to get away from the storm, too."

"Nobody in their right mind would be driving into it," Davis said.

"We are, yay!" Randolph said with buoyant sarcasm.

When the captain came back to the bridge a few minutes later, at about 3:30, Randolph alerted him to *El Yunque*'s passing.

"Yup, they speeded up," Davidson said. "They're gonna beat the storm."

"We're gonna go right into it," Davis responded.

"Seventy miles south of it," the captain said reassuringly.

4:00 P.M. to 8:00 P.M. Watch

At 4:15 P.M., shortly after chief mate Steve Shultz and AB Frank Hamm had taken over for Randolph and Davis, Hamm thought he overheard something between Davidson and Shultz. "So what did you say? There could be a chance that we could turn around?" he asked, hope creeping into his voice.

"Oh no, no, we're not going to turn around," Davidson replied. "This storm is very unpredictable."

Davidson seemed worried that the storm might follow them if they turned west, which was, in fact, the direction the storm was predicted to head.

Hamm was a big man, about six feet three inches tall and weighing close to three hundred pounds. He was easygoing and liked to make people laugh, always ready with a joke. But he was clearly not finding anything funny about the ship's course. Hamm, who lived in Jacksonville with his wife, Rochelle, was an experienced mariner. He had been at sea for decades and had seen the consequences of bad decisions up close. In 2011, he had been on the bridge of the *Horizon Producer,* another 1970s-era U.S. cargo ship, also making a Jacksonville–to–Puerto Rico run. On their way back from Puerto Rico, Hamm spotted a disabled fishing boat adrift at sea. He sounded the alarm, recalled Kurt Bruer, who was on the ship and who Hamm had just relieved from duty.

"I heard the man-overboard alarm so I went to see what was going on and they told me 'Frank just spotted these two guys,'" Bruer recalled. The big ship turned around, dropped a pilot ladder off the side, and threw the fishermen a line to pull their boat close enough so they could use the ladder. Once safely aboard the *Producer,* the fishermen told the sailors they had been blown offshore in a squall, had run out of fuel, and were taking on water. They had been drifting for three days, sharks circling them for much of the time.

So as Hamm listened to Captain Davidson talk about the storm, he knew to be worried.

Dinner in the galley that night was jerk chicken, peas, and rice. The crew had been advised that they were heading into heavy seas. Sailors in their cabins were tying down TV sets and securing favorite coffee mugs safely in their sock drawers. They packed extra pillows and life jackets under the edges of mattresses to build up the sides so they wouldn't flop out of bed if the ship rolled while they were sleeping.

Around the ship, the heavy-weather procedures were enacted. The bosun and deck crew cleared the decks of debris and dogged down the hatch covers and scuttles. The chief mate, Shultz, oversaw the crew as they checked and tightened the container locks, vehicle chains, and cargo lashings, which was made more difficult because, not expecting bad weather, they had left without extra storm lashings. The galley crew secured the dishes, kitchen tools, and supplies.

Back on the bridge a few minutes before 7:00 P.M., Davidson and Shultz were looking at the chart for nearby islands that they could duck under for cover. Shultz suggested diverting beneath the Bahamian island of San Salvador. "I agree wholeheartedly," the captain said. "Get a position."

"I'll make it so," Shultz replied.

This was not much of a course change. *El Faro* was continuing on the same course line, just sailing on the south side of the island instead of the north side.

Behind them, off *El Faro*'s stern, the sun slipped below the horizon, suffusing the sky around them in a soft light that belied what lay ahead. To the northeast, the first strands of Joaquin's long cirrus clouds were visible in the darkening sky.

It was a wistful time of day, and it triggered something in Davidson. Confiding in Shultz, he brought up his tenuous work situation. He had just received an email from TOTE Services' director of labor relations asking him to change his schedule. This was apparently so he could cover for some other staff who were going to attend classes to train for the natural gas–fueled ships that the company would be using. When the company had rejected Davidson's application to attend the training and offered him the Alaskan assignment instead, it was an insult to his already wounded pride. Perhaps it was even more ominous than that.

"I don't know what the deal is here," he told Shultz.

"When they lay this [ship] up, they're not going to take us back," the chief mate said, referring to the scheduled maintenance on *El Faro* to finalize preparations for the Alaska route.

"No, I know," Davidson replied.

"I hear what you're saying, Captain," Shultz said. "I'm in line for the chopping block."

"Yeah, same here," Davidson said.

Veterans like Davidson and Shultz rightly wondered what

kind of future they had at TOTE if the company was not training them to handle its newest and most technologically advanced
ships. The company had even asked Shultz to train an officer in
preparation for working on these ships—the same ships they ac-
tively wanted to keep him off of. Talking to his wife about it,
Shultz had noted the irony of essentially training himself out of
relevance. But she said he was never bitter about it. He just kept
doing his job.

By now it was dark out, and the wind was picking up as the
outer bands of Joaquin circled in on them.

"Let's do this," Davidson said, as if snapping out of a trance.
And with that, the two veteran sailors, each in their early fifties
and clearly worried that a changing industry was leaving them
behind, turned to the only job they had ever known and final-
ized the task at hand, altering course slightly to put a little more
distance between them and a hurricane.

"How are ya?" Jeremie Riehm said as the third mate entered
the bridge at 7:43 P.M.

"How do?" the captain replied. Shultz brought Riehm up
to speed on the latest course changes, including the decision to
use San Salvador as a windbreak. Riehm had been monitoring
the weather, and he was anxious. Some reports, like one from
weatherunderground.com, a nongovernment site that collects
data from a variety of sources around the world, were forecasting
the storm to be much more powerful than expected.

"We'll be passing clear on the backside of it," the captain said
again. "Just keep steaming. Our speed is tremendous right now."
The faster they traveled, he said, the better off they'd be. "I will
definitely be up for the better part of your watch. So if you see
anything you don't like, don't hesitate to change course and give
me a shout."

At 7:59 P.M., Captain Davidson left the bridge for the remainder of the night. From his cabin he would monitor the BVS
weather reports that came to his computer.

8:00 P.M. to Midnight Watch

Once Riehm and his helmsman, Jack Jackson, were alone, Riehm's unease surfaced almost immediately. The storm was more powerful than they thought, he told Jackson. What if it didn't follow the predicted track and head north?

"Some captains would have taken one look at that and went, 'We're going to go the Old Bahama Channel. We're not taking any chances here,'" Riehm said. Then, almost as if he was trying to convince himself, "I don't know. I'm not going to second-guess somebody. The guy's been through a lot worse than this. He's been sailing for a long, long time."

But Riehm couldn't get past the facts in front of him. They were way too close to a powerful storm, and he knew it. "Guess I'm just turning into a Chicken Little," he said. "I have a feeling like something bad is going to happen."

Jackson concurred. Hurricanes, he said, "can spin up so quickly."

At 10:53 P.M., the printer connected to the Inmarsat C terminal began automatically printing the latest weather report. "All right, this should be the one," Riehm said as he ripped off the page. He didn't like what he read. The NHC had now upgraded Joaquin to a Category 3 hurricane, with maximum sustained winds of 115 miles per hour, or 100 knots, and gusts up to 120 miles per hour.

At 11:05 Riehm called the captain in his quarters.

"Hey, Captain," he said. "Sorry to wake ya. . . . The latest weather just came in, and uhm, thought you might want to take a look at it."

Riehm read out what was on the paper.

"So at zero four hundred [4:00 A.M.], we'll be twenty-two miles from the center," he said. One option, Riehm suggested, was to head south at 2:00 A.M., likely at the Crooked Island Passage, "and that would open it up some." There was a pause while Davidson talked. "Just so you know, that's how close we'll be," Riehm said. Another pause. "You're welcome."

He hung up. "Well, he seems to think that we'll be south of it by then," Riehm said. "So the winds won't be an issue."

But the two men were looking at radically different sets of data. Even though Riehm had read Davidson the latest NHC forecast, the captain was still using his BVS report as his primary source of information. Even then, he wasn't looking at the latest BVS report, because records show he never downloaded the 11:00 P.M. package. Davidson was basing his decision on information that was twelve hours old, which meant the storm was seventy miles closer to the ship than he thought it was.

Although Riehm couldn't see them in the darkness, white cirrus clouds were likely appearing overhead, a sign of the cloud cover on the edges of the approaching storm. These initial clouds are so wispy, mariners call them "mares' tails." Gradually, as the ship steamed into the storm, the cloud formations would become more tangled, eventually becoming a dense mass of gray scud.

"I trust what he's saying," Riehm said. "It's just, being twenty miles away from hundred-knot winds, this doesn't even sound right."

"No," Jackson said. "No matter which way it's hittin' ya, it's still hundred-knot winds."

"I got a feeling," Jackson added. "Poopy suit [slang for immersion suit] and life jacket laid out."

"You got no traffic, max RPM," Riehm told Randolph as she stepped through the bridge's door with a cheery "Good morning" at 11:45 P.M. "Show ya what's going on back here in the chart room."

They both went over to the large table and leaned over the navigational chart.

"The weather report came in at twenty-three hundred," Riehm told her, meaning 11:00 P.M. As he handed it to her, he mentioned that he had told the captain about it.

Randolph examined the report for a moment, then broke out in uncontrollable laughter. "We would have been better off staying on our old track line," she said when she caught her breath.

Riehm didn't join in; perhaps he still adhered to the rigid hierarchy of the ship, ingrained into every sailor. Or maybe he just wasn't finding anything funny as he stared at the chart that had them hemmed in between islands and a storm that had ballooned in size overnight.

He continued the briefing. "At zero four hundred [4:00 A.M.] you're twenty-five miles from the center, and still going toward it," he said. "You're still diverging, zero four to zero five, instead." He was using slang for military time, meaning the ship was diverging to its farthest point from the storm from 4:00 A.M. to 4:30 A.M.

When he was done, Randolph's disbelief surfaced again. "I don't know about all that," she said, comparing the chart to the weather report. "This says maximum sustained winds of a hundred knots. Gusts to one-twenty, forty. Yeah, the whole thing's showing strengthening."

Riehm offered other options. "At zero two hundred [2:00 A.M.] you could head south instead," he said.

"This is the second time we changed our route and it just keeps heading for us," Randolph marveled.

"You know, it's eventually going to turn north," Riehm reminded her.

Randolph laughed. "*Supposed* to turn north," she said.

As the two officers went back and forth, the helmsmen considered other pressing matters.

"Getting my flashlight, lifesaver, my Gumby suit out," Jackson told Larry Davis, Randolph's helmsman, referring to his orange immersion suit, which made the wearer resemble Gumby, the Claymation TV character.

"Do you know where your EPIRB is?" Jackson asked, referring to his emergency position-indicating radio beacon.

"Ain't got no EPIRB," Davis replied.

They talked about securing the TVs in their rooms. "It ain't tied down or nothing," Davis said. "I got a feeling it's going to bite the dust."

"Man, I got everything put away in my room," Jackson said.

They chatted for a few minutes more, itemizing the tasks needed to prepare for the weather.

"Supposed to be one fifteen, with gusts up to one forty-five," Davis observed.

"Make it to Jacksonville," Jackson muttered.

"Huh?" Davis asked.

"The place to be, man, not here," Jackson replied. "Well, anyway, we're the only idiots out here."

"It's a bad deal," Davis agreed.

The wind outside picked up. The cloud deck had dropped. Dark bands of cumulonimbus clouds now thickened overhead, disgorging sheets of rain onto the ship. The wind whistled across the superstructure. With a broken anemometer, there was no way to know how fast it was moving. But it was fast.

"Blowing now like hell," Davis said.

As he prepared to leave, Riehm turned to his colleagues. "Have a last smoke," he said, then added, "All right, good night, y'all."

Chapter 7
THE SAILOR'S DILEMMA

Life at sea is regimented as a safeguard against surprise. *El Faro*'s crew had a schedule of tasks they had to stick to, routines that needed to be carried out with military-like discipline. Even if those routines were numbing and seemed endlessly predictable and rote, it was imperative that the officers check the captain's standing orders at the beginning of every shift, check the position log and chart, the compass record log, the Fathometer log. Throughout, the officer and helmsman had to maintain "situational awareness"—which Bowditch defined as the comprehension of what is known in relation to what is planned—and be prepared to adjust accordingly. The enemies of situational awareness are "complacency, ignorance, personal bias, fatigue, stress, illness, and any other condition which prevents the navigator and his team members from clearly seeing and assessing the situation."

Whatever the reason or reasons, Captain Davidson's situational awareness was compromised. He was not seeing the situation clearly. Maybe he had become complacent and over-reliant on weather technologies. Clearly he was ignorant of the time lag in the BVS reports arriving to his computer. But perhaps the pressure to perform and meet his schedule in a work environment where he felt threatened also affected his judgment.

THURSDAY, OCTOBER 1, 2015
Midnight to 4:00 A.M. Watch

After Riehm said good night and turned over the bridge, Danielle Randolph spent the next hour poring over the charts, looking for any advantage she could find as the ship sailed into the storm. The difficulty was that deviating from the track line now meant risking shallow water around the islands.

"I may have a solution," she said to her helmsman, Larry Davis, at 12:40 A.M. "We wouldn't have to worry about it until two o'clock. Our tentative position . . . gets us in a good angle, a good spot, so that we can alter course south to one eighty-six." She meant 186 degrees, almost due south. That course line, she said, would keep them five miles from any shallow water. It would also put the wind on the ship's stern, which would push them faster out of harm's way. She was plotting a course that would take the ship through the Crooked Island Passage, between Crooked Island and Long Island. This would take the ship to the Old Bahama Channel, and safety.

Then she added a caveat: "Unless this damn storm goes further south." Joaquin had been unpredictable so far. "Can't win," she said. "Every time we come further south, the storm keeps trying to follow us."

As Randolph grappled with charting alternatives that would take them away from the eye but keep the ship safe from the shallows, Davis grew incredulous. "What's he thinking?" he asked angrily, referring to the captain.

Randolph agreed. Riehm, she told Davis, had even called the captain to explain the latest weather report, and he still refused to change course.

"They said this storm could grow to a four," Davis said, observing that Joaquin was just stalled out there, spinning in place, "sucking up the hot water."

"Yeah, the ocean's been really warm," Randolph said.

"They said it's just perfect for gaining strength," Davis said. "Sure we gonna change course at two o'clock?"

"I don't know," she answered. "Gosh, I might call the captain here shortly if he doesn't come up." She paused. "I don't know if he can sleep knowing all of this."

At 1:06 A.M., Davis was looking at the radar screen, empty except for *El Faro*.

"Guess we're the only vessel in this area," he said.

Randolph just chuckled, a nervous reaction to the ludicrousness of their situation. "Yeah, all the other ships high-tailed it away," she said, barely able to get her words out. "Just us."

In the background, a weather report on satellite radio Sirius XM announced that Joaquin had been upgraded to a Category 3 hurricane.

"Oh my God," Randolph said.

"This is fixing to get interesting," Davis said.

"Miiissstaaake," Randolph replied, drawing out the word.

"Can't pound your way through them waves," Davis observed. "Break the ship in half."

"We're going to hurry and get out of it. That's what *El Yunque* did," Randolph noted.

"They knew they were going to get out of it," Davis replied. "We [are] going *into* it."

"I'm going to give the captain a call and see if he wants to come up," Randolph said.

At 1:20 A.M. she called the captain, asleep in his bunk.

"It isn't looking good," she told him. She wanted to alter route and head straight south, through the passage.

"He said to run it," Randolph announced when she hung up, meaning the captain had denied her request to change their route. She laughed again. "Hold on to your ass!"

"So, we're going to stay on this course?" Davis asked, in disbelief.

A minute later, she noted that "the captain was sound asleep. It took a few rings for him to answer."

And here was the sailor's dilemma. Everything about their training drilled into them the importance of following orders. On a ship, even a civilian ship, "orders are given and expected to be followed down the chain of command without hesitation or question," according to Bowditch. Anarchy on a ship at sea was a fast route to disaster. Earlier, the chief mate had recounted a story about a second mate he had heard about who had altered a ship's course without notifying the captain just to see the other side of an island. The act was so far out of the bounds of reason that "you can't even get mad at a guy like that," Davidson had said. "You gotta look at them and go 'Really?'"

But what does a crew do when it becomes increasingly certain—and unanimous—that the captain's orders are putting them all in danger? Even Bowditch acknowledges that in perilous situations it can become necessary to consult with other members of the team and challenge a captain's order: "Many serious groundings could have been prevented by the simple exchange of information from crew to captain." By this point, both Riehm and Randolph had explicitly told the captain about their unease, and Davidson had ignored them, sticking to his plan. How far should the officers take their concerns? Both were younger than the captain. Less confident. Less experienced in heavy weather.

And the rules of the ship are so ingrained that they are hard to break, even in the darkest of circumstances. Maritime history is rife with examples of captains who set course for disaster, ignoring crew members whose tepid warnings went unheeded. In 1873 the captain of a passenger ship entering a harbor in Nova Scotia didn't consult the charts, misinterpreted a shore light, and went to bed. The ship crashed and more than five hundred people died. In 1912, the captain of the *Titanic* believed his ship was unsinkable and sailed into a field of icebergs. An iceberg ruptured five watertight compartments. Engineers had presumed no accident would affect more than three at one time.

"If one person has unquestioned, absolute authority over a system, a human error by that person will not be checked by others," the sociologist Charles Perrow wrote about shipmasters in his influential study *Normal Accidents,* which analyzed industrial accidents in a variety of fields, including marine transportation.

It is impossible to know what had hardened Captain Davidson's resolve to maintain the ship's course and ignore his officers. His time in the rough waters of the Pacific Northwest might have caused him to underestimate an Atlantic storm like Joaquin. But there is evidence he simply lowered his guard, and relied too heavily on technology. This is a danger for modern shipmasters. Technology—GPS, autopilot, sensors—runs the ship. Some marine academies have stopped teaching celestial navigation because of the widespread use of GPS. The concern is that an overreliance on these systems will lessen the sailor's connection to the sea, and diminish the respect it warrants.

Captains like the *Falcon Arrow*'s Gokhale, who had repeatedly sailed the uncontained ocean, where there are no islands to act as windbreaks (compared to the mostly coastal routes that made up the bulk of Davidson's career), still knew enough to be humbled by the vast power of the weather and water.

After talking with Davidson, Randolph and Davis adjusted their course slightly, to 116 degrees south, to the waypoints the captain and chief mate had earlier logged. At 1:24 A.M. the ship was steaming past Rum Cay and San Salvador Island. A few minutes later, the wind picked up and the boat started to heel. "Starting to hear the wind now," Randolph said.

At *El Faro*'s position, the wind was blowing about 60 miles per hour and the seas were about fourteen feet. The waves were heaping up, with white foam breaking off from the peaks and blowing with the wind.

The two sailors noted the increasing intensity of the weather with a mixture of alarm and the stoicism of seasoned sailors. Lightning flashed off the bow, big swells lifted the boat. The wind outside drummed the air, like thousands of horse hooves galloping across the sky. "Whee! Look at that spray!" Randolph exclaimed at 2:42 A.M., as a huge plume of water erupted over them.

"Figured the captain would be here," Davis mused.

"I thought so, too. I'm surprised," Randolph answered. "He said he would probably be up here."

"He'll play hero tomorrow," Davis said.

Randolph laughed, this time at the humor.

Big swells continued to pitch the ship as the wind whined through the superstructure. The two sailors heard thumps and crashes from the decks below as loose objects fell. The ship was on autopilot, nicknamed the Iron Mike. On calm seas it steered the ship along a programmed route. If the seas were rough and pushed the bow off-route, the off-course alarm would sound. Right now, that alarm was beeping repeatedly.

"We're just breaking her in and getting it ready for Alaska," Randolph said. A few minutes later she added, "If this ship can't handle this storm, *suuure* as hell can't handle Alaska," she said, drawing out the word for emphasis.

And then she seemed to revise her earlier criticism of *El Faro* as a rust bucket. "The only comforting thing is that this ship has much better steel than the *El Morro*," she said, in what was perhaps a bout of wishful thinking. "She's still pretty solid."

Above them on the flying bridge came the sound of metal crashing. "Rhut-rho," Randolph said, doing her best Scooby-

Doo imitation. The plan at that point was to power through the weather and emerge on the other side in about an hour with the storm behind them. The waves kept pushing the bow of the ship around, and every few minutes the off-course alarm would sound. At 3:22 A.M. Randolph said, simply, "Hello, Joaquin."

"Look at that radar," Davis marveled a few minutes later, as the storm's splotchy cloud cover filled the screen.

"It's just getting bigger," Randolph said. "Our path is going right through it."

Randolph couldn't have known it at the time, but she was more correct than she realized. The ship wasn't heading south of the storm, as Captain Davidson believed. It was heading *north* of the eye—but just barely.

4:00 A.M. to 8:00 A.M. Watch

El Faro was alone on the radar, steaming south into a wind-whipped sea, when chief mate Steve Shultz arrived at 3:45 A.M. to prepare for his shift.

Randolph and Davis explained their course to him, adding that there was no traffic and zero visibility. Before they finished the brief, the phone rang. Howard Schoenly, the second assistant engineer, who was on shift in the engine room, wanted to let the bridge know that he would be doing a regularly scheduled maintenance procedure to clean the boiler tubes of mineral buildup, a process known as blowing tubes. Schoenly's call suggests that at least some members of the crew believed everything was proceeding normally at this point. The drama on the bridge, the officers' concerns, the chart changes, possibly even the latest weather information—apparently none of these developments had made their way down to the engine room.

In any case, blowing tubes meant that the ship's speed would be reduced for a period of time.

As Randolph talked with the engine room, Shultz moved over

to helmsman Davis. Shultz's helmsman, Frank Hamm, hadn't arrived on the bridge yet.

"So you can't see a thing, huh?" Shultz asked.

"Yeah," Davis answered. "If anybody's out there, they gotta be a damn fool."

After hanging up with Schoenly, Randolph continued briefing Shultz. The ship was holding course, she said, although with a big wave or a steep pitch "she might lose it a little bit." As if on cue, the steering alarm went off. "She'll come back," Randolph said. "She did that a couple of times just because we pitched so bad."

With that, Randolph told Shultz to have a good watch and went down to her cabin, where she promptly flipped open her laptop and composed an email to her mother. "Not sure if you have been following the weather at all," she wrote, "but there is a hurrican [*sic*] out here and we are heading straight into it. Category 3 last we checked. Winds are super bad and seas are not great. Love to everyone." She hit send at 4:39 A.M.

When Laurie Bobillot saw the message the next day, she knew with a mother's intuition that something was wrong. Her kid was scared, and Danielle Randolph didn't scare easily. For one thing, Danielle never signed off "Love to everyone." That was not her style. Typically she'd just put her first initial, or write "Say hi to everyone." If her daughter was scared, then Bobillot knew to be scared. It was almost as if Danielle was saying goodbye.

By now Joaquin was bearing down on Samana Cay. The storm was holding steady as a Category 3 hurricane, with 120-mile-per-hour winds that encircled the eye in a thirty-five-mile radius. While the captain was under the impression that his ship had sailed south and east of the storm, in reality *El Faro* was about sixty nautical miles northwest of Joaquin's twisting heart.

The winds alone were a powerfully dangerous force to reckon with. But what they were doing to the water was a tremendous hazard for the ship.

At this point the core of Joaquin was a well-formed and terrible marvel of nature, awesome in scale and force. It was as if an enraged angel had constructed a cathedral made of pure power. Joaquin's cloud deck extended into the upper troposphere, eleven miles above the earth, where the temperatures are less than minus 50 degrees Fahrenheit and the moisture from the seawater being sucked into the central funnel freezes into sheets of ice. Climate scientist Kerry Emanuel has likened the center of the storm to being inside a twenty-mile-wide Roman coliseum that was ten miles high, with "a cascade of ice crystals falling along the coliseum's blinding white walls." The eye wall itself was made of thick, dense clouds, above which a canopy of tightly packed cumulonimbus clouds spread in spiraling striated bands of billowing gray that extended for miles and miles, each band a thunderstorm system in itself. Feathery cirrus clouds spun off at the storm's outer edges.

El Faro was steaming toward a near perfect intersection with this colossus of raw energy.

At about 3:50 A.M., Hamm entered the bridge. He made some coffee with the chief mate and then Shultz and Hamm reviewed the chart course amid the bleating of the steering alarm, which was so frequent that Shultz simply turned it off. Shultz gave some commands to keep the ship on course, but the waves kept knocking her off. "How much longer of this?" Hamm asked.

"Hours," Shultz replied.

At 4:09 A.M., Captain Davidson finally stepped onto the bridge.

"Hello there, Captain," Shultz greeted him.

"There's nothing bad about this ride," Davidson responded. Apparently he felt rested. "Sleeping like a baby," he said.

Perhaps Davidson sensed his crew's unease, and as captain he wanted to reassure them with his confidence. Perhaps he genuinely believed what he said. Either way, his manner seemed untethered from the reality around them.

"Well, this is every day in Alaska," he said. "This is what it's like."

"That's what I said when I walked up here," Shultz echoed. "I said this is uh—this is every day in Alaska."

"A typical winter day in Alaska," Davidson said again, now sounding as repetitive as the steering alarm. He continued: "I mean, we're not even rolling. We're not even pitching. We're not pounding." The captain had not yet looked at the weather reports on the bridge.

"You on the Mike?" he asked, referring to the Iron Mike.

"I turned off the course alarm," Shultz said. "It was going off every five seconds."

At 4:12 A.M., Shultz made mention of a slight tilt in the ship to the starboard side. Shultz believed it was wind heeling the ship over, not a list caused by an imbalance in the ship's weight.

"Cap'n, yeah, we're goin' like this," Shultz said, indicating the tilt. "I'm guessing it's on the port bow and . . ."

"Port side, yeah," Davidson concurred.

"Wind," Shultz added, reiterating his belief that the ship was heeling starboard.

"The only way to do a counter on this is to fill the port-side ramp tank up," Davidson said. Filling the port ballast tank would make that side of the ship heavier and weigh it down. He didn't give the order, though. Counterbalancing by shifting ballast was not something to be done lightly. It required knowing the precise weight distribution on the ship, including whether any cargo had shifted, to understand how the vessel's stability would be affected by the adjustment. If there was a flooding issue, for instance, the water could rush to the other side with dangerous momentum.

"Oh, it's howling out there," the captain observed.

"Can't tell the direction," Shultz said, meaning the wind, due to the broken anemometer. "But our forecast had it coming around to starboard."

Shultz was confused because he had noticed the winds were on the port side, and that's why they had what he thought was a starboard heel. This didn't make sense, at least not according to their understanding of where they were in relation to the storm. They thought they were on the south side of Joaquin, where the winds would be hitting the starboard side of the ship.

"The wind?" Davidson asked, always confident. "It will [come around] eventually."

At 4:34 A.M., Davidson ducked out of the bridge, saying he was heading down to the galley to see how they were making out. Ten minutes after the captain left, the phone rang. Shultz picked it up. "Okay, I understand you," he said into the receiver. He hung up and immediately rang belowdecks.

"Captain, chief mate. The chief engineer just called," Shultz said. "Something about the list and the oil levels."

El Faro's oil was held in a reservoir and sucked up through a flared-end pipe called a bellmouth pump to circulate through the engine. If the ship listed too severely to either side, it could pull the oil away from the pump. Without oil, the engine would seize, so built-in safety mechanisms automatically shut the engine down before this happened. Any list was a danger, but in *El Faro*'s engine design, the pump was offset from the center 22 inches to the starboard, so a port list was more risky. It's not clear whether the ship's engineers knew the pipe was offset.

Additionally, the ship had left port without fully filling the oil reservoir. *El Faro*'s operating oil levels ranged from a low of 18 inches to a high of 33 inches. When the ship was expecting bad weather, some TOTE engineers liked to fill the oil reservoir up a little beyond the high capacity. But when *El Faro* left Jacksonville on September 29, the oil levels were only at 24 inches. Because of the 22-inch pump offset, this meant a fifteen-degree list to port would automatically shut down the engine.

The captain hurried back to the bridge and called the chief engineer, Richard Pusatere, who told him the ship needed to ease off the list.

"Wants to take it off the list, so let's put it in hand steering," Davidson said, indicating that the helmsman should take the ship off the Iron Mike and steer manually. Shultz gave the order to Hamm.

"The sumps are acting up," Davidson said. "To be expected."

A minute later, the captain noticed that the barometric pressure, which measured the weight of the atmosphere above the ship, was now reading 960 millibars. The average barometric pressure at sea level is about 1,013 millibars. But in a hurricane, the rising heat causes the pressure at sea level to drop. The ship's reading showed just how close it was to the eye of Joaquin, which had barometric pressure of 948 millibars at 5:00 A.M. The ship was rocking, causing Hamm to yell out "Whooo!" in alarm. Davidson took over the steering commands. "Master has the conn," he said, and began giving orders to the helmsman. "Go left ten . . . midship . . . right twenty . . . hold that course."

Hamm was visibly tense, trying to hand steer the ship through Joaquin. "Stand up straight and relax," Davidson told him.

"I'm relaxed, Captain," Hamm replied.

After a few minutes, Davidson announced that he was going down to his cabin to see if there were any messages. "I'll be back up," he said.

And that's when Davidson finally checked the bridge's weather reports against the BVS reports he was getting in his cabin's office. He came rushing back to the bridge four minutes later. "Mate, is this synced up with the most recent broadcast?" he asked Shultz.

"Yes, sir," Shultz replied. "I saved the five o'clock real quick so I could plot it once you got up here."

"Here's the thing," Davidson said. "We're getting conflicting reports as to where the center of the storm is."

It was 5:02 A.M. This was the captain's first indication that the information he'd used to plot the course—information he'd stood by through the night, information his officers had started questioning at eleven o'clock the night before—was off by at least six hours, which would mean they were dangerously close to the eye of Joaquin. Yet, it was unclear that he understood the mistake. He asked his chief mate to call up the weather again, and then gave some steering commands to get them off the list.

"This is what it's like every day in Alaska," Davidson said again at 5:10 A.M. Then a little later, "It's only going to get better from here."

He was dead wrong. The barometric pressure had now dropped to 950 millibars. At that point, the center of the storm was only forty miles away, almost directly in front of them. They were sailing straight into the northern edge of the hurricane's eye wall, the dangerous semicircle.

At 5:43 A.M. the engine room called again. Davidson picked up and listened. "We've got a problem," he announced. He drew the word out so that it sounded like "prrroooblemm." Davidson turned to the chief mate. Engineers had discovered water flooding through an open scuttle into Hold 3, one of the cargo storage areas in the center of the ship, just in front of the engine room. This was what was causing the list. How exactly this breach happened isn't known. The heavy-weather preparations the night before would have included dogging all watertight doors and hatches. But anything can happen in a storm. Equipment left out by the welders could have slammed into the hatch cover. Or maybe the scuttle wasn't watertight to begin with. Able seaman Kurt Bruer remembered a scuttle midship on the main deck that couldn't be closed because of corrosion. "That particular scuttle, they could never get it repaired," he recalled. "It was always left open."

"Watch your step," the captain instructed Shultz. "Go down to three hold. Start the pumping right now."

Davidson wanted an experienced hand down there; that's why he chose his chief mate. He also wanted Shultz to wake up one of the third assistant engineers to take with him. "It's unsafe to go down in the cargo hold with gear adrift," the captain reminded him. Cargo afloat in the hold could pin a man or crush him. Anyone entering the hold had to be careful.

Before Shultz could leave, the phone rang again and he answered it.

"Go ahead," Shultz said. "Okay. Are we able to pump the bilges?"

The captain quickly took the phone from Shultz.

"Bilge pump running, water rising. Okay. Can we pump from the starboard ramp tanks to port?" Davidson asked. "Let's do that."

The water was coming into the hold faster than it could be pumped out, and the ship couldn't recover from the starboard list. The engineers were worried about the oil levels dropping and shutting off the engine if the ship was not stabilized. *El Faro* used ballast tanks filled with metal. It also had two smaller, liquid-filled ramp tanks, which were used to correct lists less than two degrees.

"Okay, what I'm going to do, I'm going to turn the ship and get the wind on the north side," the captain said into the phone. "Give us a port list."

Davidson was hoping the ramp tank transfer, combined with putting the starboard side to the wind, would correct the list. This was a drastic move—one the captain would not have resorted to unless he realized the danger of the situation. Shipmasters spend a lot of time studying ship stability and how to maintain it. Davidson would have known that trying to accomplish a windward heel in a storm to counteract a list caused by flooding could dangerously shift the water. But he must have known he had to act quickly or the oil would lose contact with the engine and possibly bring the ship to a stop.

He gave steering commands to the helmsman: "Put your rudder left twenty . . . midship . . . three five zero." Then he called back down to the engine room. "All right, we've got a nice port list. Can you stop transferring?"

The report back was initially good. "We got the water blown out," the engineer on the phone relayed. "The worst area's already dried out."

At 5:59 A.M. Danielle Randolph appeared on the bridge. She couldn't sleep. "Hi," Davidson said brightly.

"How are you, Captain?" she asked.

"A scuttle popped open and there's a little bit of water in three hold, and they're pumping it out right now," he told her.

Just then, the bridge phone rang again. "Captain," Davidson answered. "All through the ventilation?"

"Want me to bring it back over to starboard?" he asked. "All right."

The earlier call, reporting success drying out the hold, was premature. The ship continued to list, and the severe angle was allowing water to flood in through the main ventilation shafts along the ship's freeboard deck, along the upper edge of the hull, and down into the hold. Davidson gave commands for getting the ship's starboard side to the wind and transferring ballast back to port. Then he turned to Randolph. "Hey, Second Mate," he said. "See what's up with this radar right here." The radar had stopped working.

Chief Mate Shultz returned from Hold 3. The scuttles were half-open, he said. They had indeed been checked last night, but something had forced them open since. It could have been gear slamming around on deck, or the waves torquing the old ship's hull. In any case, he reported, the guys had closed it, "so we're good." He thought he had fixed and secured the main source of flooding. But he didn't know about the other major source of flooding, the water crashing onto the deck and streaming through the ventilation shafts along the edge of the freeboard deck. These were always left open to allow fumes from below decks, where the cars were stored, to escape. As a result, there was still a lot of water in Hold 3, and the list had not gotten any better. Davidson asked him to go back down with a radio. "We need ears and eyes down there," he said.

They were now entering the inner bands of the storm, about twenty miles northwest of Joaquin's eye. One-hundred-and-twenty-mile-per-hour winds screamed over the ship, making a deep, layered thrumming sound. It was still too dark to see outside, but the air was thick with sea foam whipped up by the winds. There may have been moments when they descended a big crest and could glimpse the ocean's surface, which was completely white with driving spray. Waves higher than fifty feet, some as high as the ship itself and weighing two thousand pounds per cubic meter, were slamming against her hull.

There was the sound of crashing outside. "There goes the lawn furniture," Randolph quipped. Then the second mate, who had the radar working again, turned to the captain. "If you don't need me out, you want me to stay with you?" she asked.

"Please," the captain said. "It's just the . . . it's just the . . ." He didn't finish.

The chief mate radioed from below: "Chief mate to bridge."

"Go ahead, Chief Mate," the captain said.

Shultz told him they needed to transfer water from the port to the starboard ramp tank. The captain gave the okay and hung up.

"I'm not liking this list," Davidson said to no one in particular.

Moments later, the low vibration of the ship's engine suddenly went quiet. The rumbling of the engine was one of the constants of life at sea, and its cessation created an unsettling void that was immediately replaced by dread. "I think we lost the plant," Davidson said. The switch to a port list had ultimately moved the oil away from the pump, causing the engine to shut down. It was 6:13 A.M.

El Faro was now without propulsion near the center of a Category 3 hurricane. The ship's bow was pointed north, and she was drifting amid thirty-foot seas, which was the average of the highest one-third of the waves. The waves would now gradually take control of the ship's direction and start turning her little by little until she was broadside to the wind. Then the wind and waves together would drag her on a drift. Davidson called down to the engine room. After a quick conversation, he said, "They'll bring everything back up online."

Within minutes, the engine room rang again. Water was coming in through the ventilation shafts down there.

Davidson asked Randolph to go down and wake Third Mate Riehm. This was escalating to an all-hands-in-the-bridge situation. Randolph asked if she could also dash to her cabin and change into her work clothes. The captain gave the okay.

Eight minutes later, Riehm arrived in the bridge. "Am I relieving watch?" he asked, flustered. "Tell me what to do and I'll do it."

The captain was busy on the radio to the engine room. "They're getting that boiler back up. They're getting that lube oil pressure back up," he said.

But all evidence indicates that the boiler had never shut down. It was solely the pump losing contact with the oil that had triggered the automatic engine shutdown.

Meanwhile, the list continued, and more water was reported from below. *El Faro*'s bow now pointed northeast as the ship drifted south, pushed by the storm. Davidson ordered Randolph to compose a message. Not necessarily an emergency—more like a notification that the ship was in distress. Davidson instructed her to compose emails to send to TOTE and to the U.S. Coast Guard alerting them of their situation. Randolph composed aloud, mentioning flooding cargo, the time, and the ship's position.

"Don't send anything yet," Davidson said at 6:39 A.M., perhaps hoping against all available evidence that the ship would

regain power and they would be able to steam out of their pre-
dicament, saving him the embarrassment of contacting management.

Even as they drifted, the captain gave Hamm steering orders
to continue trying to correct the list. "You got some turns now,"
he told Hamm at 6:44 A.M. "Keep your rudder . . . right twenty
right now."

As the ship responded to the maneuver, Davidson, looking for
any good news he could cling to, proclaimed, "That's a small
victory right there."

By now, the ship's list was forcing the sailors to lean their
weight forward to compensate. At the helm, Hamm grunted
with the effort. "You okay?" Davidson asked.

"You're getting a leg workout," Randolph teased. "Feeling
those thighs burn?"

From outside came the sound of repeated thuds as bits of the
ship were ripped off by the howling wind and the raging waves.

"That's why I don't go out there," Davidson said. "That's a
piece of the handrail, right?"

"Yup," Randolph replied. Then she took it upon herself to
make coffee for everyone—an effort, perhaps, to introduce a
calming bit of normalcy as each minute grew more and more
fraught.

"Cream and sugar?" she asked the room.

"Hook it up," Hamm said. "Do your thing."

Randolph laughed.

"Sugar is fine with the captain, right?" she asked.

But Davidson was distracted.

"Give me the Splenda, not the regular sugar," Hamm requested.

"I'm going to be right back up, okay?" Davidson said, then he
ducked down to his office. He was back on the bridge in two
minutes. At 6:54 A.M., he was on the phone with someone, it's
not clear who, giving a status update.

"It's miserable right now," he said. "We got all the wind on the starboard side here. Now a scuttle was left open or popped open or whatever, so we got some flooding down in three hold—a significant amount. Everybody's safe right now, we're not going to abandon ship. We're going to stay with the ship. We are in dire straits right now. Okay, I'm going to call the office and tell them. There's no need to ring the general alarm yet. We're not abandoning ship. The engineers are trying to get the plant back. So we're working on it, okay?"

When Davidson hung up the phone, Hamm asked how long the weather was going to last.

"Should get better all the time," the captain mused as the storm wailed outside the window. "Right now we're on the backside of it."

Davidson still did not comprehend the ship's position in relation to the storm. The ship was north of the storm's center, and Joaquin's winds were pushing *El Faro* south, toward the eye, which was now stalled about fifteen miles above Samana Cay.

Outside, more crashing sounds as the wind continued to claw at the ship, ripping off bits and pieces and flinging them into the sea.

At 6:56 A.M., Davidson called the chief engineer for a status check. "I understand," he said into the receiver. "Look, what I'm going to do is, I'm going to call the office and let them know we have a situation here, and we will go from there."

Randolph could guess what the problem was, listening to the captain's conversation. "They having trouble getting it back online?" she asked, referring to the engine.

"Yeah, because of the list," Davidson replied. He picked up the satellite phone that was a direct link to TOTE's onshore support services. Captain John Lawrence was the designated person ashore, or DPA. It was 6:59 A.M. No answer. Davidson had to leave a message.

"Captain Lawrence, Captain Davidson, Thursday morning

zero seven hundred. We have a navigational incident. I'll keep it short. A scuttle popped open on two deck and we were having/ had some free communication of water go down the three hold. Have/getting a pretty good list. I want to, uh, just touch . . . contact you verbally here. Everybody's safe, yeah, but I want to talk to you."

Within seconds the phone rang back. It wasn't the DPA; it was an operator. "Yes, this is a marine emergency and I need to also notify a QI. Are you connecting me through to a QI?" Davidson was getting frustrated. "QI" referred to a qualified individual, another term for the DPA. "Yes, ma'am, my name is Michael Davidson . . . ship's master . . . *El Faro*." His frustration was growing. "The clock is ticking. May I please speak with a QI?"

But instead of connecting him, the operator asked him to spell the name of the ship. "Echo Lima space Foxtrot Alpha Romeo Oscar," Davidson recited through what must have been clenched teeth.

When he was finished, he gave her the phone number and ship's position. She put him on hold, so he used the time to get back on the radio for a status update on the water flooding the hold. "We're leaning pretty good over to port," Shultz reported.

"Can you tell me if it's decreasing or increasing?" the captain asked.

"I can't tell, Captain," Shultz replied. "Seems as if it's going down."

Then Davidson was back on the satellite phone arguing with the operator.

"I have a marine emergency and I would like to speak with a QI," he said at 7:05 A.M. "We had a hull breach—a scuttle blew open during a storm. We have water down in three hold. We have a heavy list. We've lost the main propulsion unit. The engineers cannot get it going. Can I speak with a QI, please?"

Finally, at 7:07 A.M., a full seven minutes after he first got on the phone, he was put through to Captain Lawrence. But then, confronted with a corporate representative, Davidson swallowed his frustration, his panic—whatever understanding and acceptance of the urgency of their situation he had—and instead relayed a message in the calm, measured tones of someone giving a midday report to the office.

"I'm real good," he said, inexplicably. "We have secured the source of the water coming into the vessel. A scuttle was blown open by the force of the water perhaps, no one knows. Can't tell. It's since been closed. However, three hold's got considerable amount of water in it. We have a very, very healthy port list. The engineers cannot get lube oil pressure on the plant, therefore we've got no main engine. And let me give you a latitude and longitude. I just wanted to give you a heads-up before I push that—push that button."

For an agonizing five more minutes he talked with Lawrence, relaying the ship's latitude and longitude, giving reassurances that the crew was safe, that the priority was to stay with the ship because the weather was "ferocious out there," providing details on the seas and visibility and the degree of the ship's list ("ten to fifteen degrees, but a lot of that is wind heel"), and reiterating that the engine room was pumping the water out of the hold.

Finally, at 7:12 A.M. he said, "What I wanted to do, I want to push that SSAS [ship security alert system] button. I want to send some alarms in our GMDSS console. I want to wake everybody up."

Captain Michael Davidson, a merchant mariner his whole life, a captain for more than ten years, was asking permission in an emergency situation to ring the alarms that might save his crew.

"Okay," he continued. "I just wanted to give you that courtesy so you wouldn't be blindsided by it and have the opportunity . . . Everybody's safe right now. We're in survival mode now. . . . Yup, thank you, sir."

He hung up. Then, back in a command frame of mind, he

turned to Danielle Randolph. "All right, Second Mate," he said. "Send that message."

The latitude he gave Lawrence was 23.48 degrees north, 73.86 degrees west.

The 8:00 A.M. NHC report put the center of Joaquin at 23.2 degrees north by 73.7 degrees west.

Chapter 8

THE SECOND EYE

Amid the mirrored glass high-rises of Miami's Brickell Avenue, the functional concrete of the U.S. Coast Guard's District 7 headquarters stands out for its drabness. The downtown location posed a unique challenge for the sailors, pilots, and air crews accustomed to wide-open spaces. Matthew Chancery, for instance, a thirty-four-year-old operations unit controller, worked in a windowless room behind a secure door in the building—D7's Command Center. In the past, prior to transferring to District 7, he had been a cutterman, assigned to ships patrolling the rolling waters of the Atlantic and Caribbean.

But Chancery's current job gave him a view far wider than the horizon from a ship's rail. When he was on watch, as he was on the morning of Thursday, October 1, 2015, the entirety of the district's 1.9-million-nautical-square-mile area of responsibility spread out before him on a series of linked computer screens.

"On the cutter I did the mission," the petty officer, whose face is round and boyishly handsome, liked to say. "Here I control the mission."

The Command Center is the cerebral cortex for District 7's main search and rescue and law enforcement functions; they are responsible for water stretching north of Bermuda and south almost to the Caribbean. In reality, District 7's watchers keep an eye on the waters all the way down to Venezuela. It's a bustling expanse, encompassing busy commercial shipping lanes, one of the most populous pleasure boat zones in the world, and the majority of seabound drug-smuggling routes from South America. A sign by the Command Center's door reads ¼ OF ALL U.S. COAST GUARD SEARCH AND RESCUE AND ⅔ OF ALL U.S. COAST GUARD LAW ENFORCEMENT IS HANDLED BY THE SEVENTH DISTRICT COMMAND CENTER.

Chancery's watch started at 6:00 A.M., but he arrived a few minutes early for the "pass down," a run-through of priority information from the previous shift. Then he took his seat at the black search and rescue controller desk, where a triptych of computer screens partially encircled him. His computer was programmed to receive EPIRBs, the radio beacon for vessels; ELTs, or emergency location transmitters, the beacon for aircraft; and PLBs, personal locator beacons, which locate individuals. Another computer was set up to receive emergency emails sent by ships at sea. In front of him was the "Wall of Knowledge," twelve sixty-five-inch TV screens that could each broadcast separate images or be linked to display one outsize image.

At 7:33 A.M., Chancery was on his second cup of coffee when his computer pinged with an email from the Coast Guard's Atlantic Area Command in Portsmouth, Virginia.

The email reported that the Atlantic area had just received an Inmarsat C distress signal from the steamship *El Faro*. There was no immediate danger of sinking, according to the communication from the ship, but Atlantic Area wanted District 7 to assume search and rescue command coordination, which included reaching out to Bahamian authorities to see if they had any assets in the area.

The first thing Chancery did was try to find a number for the ship's satellite phone. He wanted to get information firsthand. He found the TOTE contact number on the Atlantic Area Command's email and picked up the phone. John Lawrence answered.

"Hey, John, this is Petty Officer Chancery. I'm from the Coast Guard in Miami, Florida. How are you?"

"Yes, sir."

"You were listed as a POC for the *El Faro*."

"That's correct," Lawrence said.

"Okay, do you have contact or direct communications with the vessel?" Chancery asked.

"I did," a nervous Lawrence replied. "They had called me. I was just actually trying to call them back, and I couldn't—the satellite is dropping the call. I can give you the phone number."

"Yeah, give me the phone number for the vessel," Chancery said. "That's fine."

Lawrence read out the number and Chancery took it down.

"Can you tell me what the, what the plans, what you [are] planning on doing now?" Lawrence asked haltingly.

"So right now," Chancery answered, "based off all the information that you've provided me—I'm not in the distress phase currently."

"Okay," Lawrence said.

"Because they're not at risk of sinking and they have dewatered," Chancery continued. "I'm looking at—they are without power and engines?"

"Correct," Lawrence replied.

"So are they able to anchor that boat right there?" the guardsman asked.

"They're forty-eight miles east of San Salvador, so I don't think so," Lawrence replied.

"The position that I'm looking at, they should be able to anchor," Chancery countered.

According to the charts Chancery was consulting, the water at *El Faro*'s location appeared shallow enough for the ship to drop anchor and wait out the storm. Chancery told Lawrence that if the ship were simply disabled and not in distress, it would be the company's responsibility to provide tugboat assistance.

"All right, I'm going to try to give the ship a call and get a better handle on what the situation is and what's going on now," Chancery said. "If you hear from them just give me a call back."

As they were finishing their conversation, an alert went off on Chancery's computer. "Attention, emergency," a computer voice announced. It was *El Faro*'s EPIRB.

Now Chancery was alarmed. He quickly hung up and checked the information. The EPIRB, frustratingly, did not give a location—just where it was registered. *El Faro*'s EPIRB was not encoded with GPS. This was another of the safety regulations the ship was not required to comply with due to its age. The last position given by Lawrence put the ship right next to the eye of Joaquin.

Chancery immediately tried calling *El Faro*'s satellite phone while attempting to locate it via satellite images and the ship's automatic identification system (AIS), a vessel-tracking program. He couldn't get through. He tried again. Over the next several minutes he tried repeatedly, hoping the lack of connection was just a poorly positioned satellite.

Chancery briefed the search and rescue mission coordinator, Captain Todd Coggeshall, that he might have a case. Then he contacted Bahamian authorities to see if they had any assets nearby, only to learn that the Royal Bahamas Defence Force ships had fled to Key West to wait out the storm. Next Chancery checked the maps for Coast Guard resources. There was the helicopter base on Great Inagua. He called the island and asked if a helo could fly over to check out *El Faro*'s location. The duty officer looked into it, but replied that the helicopters were "out of parameters." This meant that the wind speeds were too high for the helicopter's blades to turn without becoming unstable and destroying the rotors. Even if they managed to get the helo airborne, it would be extremely dangerous flying into vortex winds so close to the eye.

Chancery checked in again with Coggeshall, who recommended calling the National Hurricane Center and the Keesler Air Force Base in Mississippi to ask any hurricane hunters flying over the storm to please call out over their radios in an attempt to hail the ship.

Meanwhile, Chancery looked for other ships in the area on a computer screen using *El Faro*'s AIS information. At first it appeared there were none, a completely empty stretch of ocean . . . but then he noticed a blip on the screen. Astonishingly, the *Emerald Express,* a Panamanian-flagged 173-foot general cargo ship out of Fort Lauderdale, was nearby. Chancery called the shipmaster. The *Emerald Express* was sailing south at the time, desperately battling the wind and waves, trying to find some protection behind one of the islands. The ship's captain put it bluntly, according to Coggeshall, telling him, "I'm more than willing to make callouts for you, but there is no way I'm going over there." The ship, which had been about twenty-five nautical miles behind *El Faro,* ended up sailing south of Crooked Island for lee protection. It eventually got blown more than ten miles inland, down a mangrove swamp, where it would remain, stranded, for months.

The men in the Command Center were out of options. All they could do now was wait, prepare—and hope to hear back from *El Faro.*

Roughly 160 miles southwest of *El Faro*'s last known position, the sky was still blue on the morning of Thursday, October 1,

with patches of clouds here and there. Captain Gelera was working the 6:00 A.M. to noon shift on the bridge of the *Minouche*,
motoring through the fourth day of a smooth, uneventful journey. But as the morning wore on, thicker, darker clusters of
clouds began to stain the sky to the north, and the seas picked
up.

Waves started rocking the ship, pitching her forward, rolling
her from side to side.

"Ship running normally at 7–8 kts," Gelera wrote in the ship's
log. "Sea from moderate to becoming rough, w/ w'ly heavy
swells and w'ly strong wind with gust. Barometer 101."

Still, the weather report for the stretch of ocean between their
current location and their destination looked fine—nothing to
worry about.

At 11:00 A.M., Gelera checked the NHC's latest forecast as it
came in on the Inmarsat C terminal. Joaquin was now less than
200 miles away, over Samana Cay, and moving southwest toward
the *Minouche* at about six miles an hour. The winds around the
eye were blowing at 125 miles per hour, Category 3 strength, in
a 45-mile radius. Tropical storm–force winds of up to 73 miles
per hour were blowing in a 140-mile radius. The storm had not
veered due west, as forecasters had earlier predicted.

As the *Minouche* puttered along at about seven knots, Cuba's
green, rocky coast loomed up off the starboard side: Cayo Ro-
mano, Playa Santa Lucia, Puerto Padre. Soon the ship would
pass Punta de Maisí on the extreme eastern tip of the island. At
that point they would enter the Windward Passage. Across the
pass was Haiti. Gelera expected to be anchored off the wharf at
Port-de-Paix by nightfall. It would be a relief to get off these
choppy waters.

As the day progressed the seas grew rougher. The hurricane
was descending south, getting closer. It was also growing stronger. By noon Joaquin had swelled to a Category 4 storm, blowing more than 130 miles per hour. A veil of cirrostratus clouds

gradually covered the sky, as the outer bands of Joaquin crept closer to the *Minouche*.

Before ending his shift, Gelera sent a routine report to the ship's agent in Miami, reporting the conditions and that they were proceeding on course. Then he logged the weather: "Very rough seas w/ heavy swell. V/L [vessel] pitching and rolling heavily. All are OK/ Normal for the vessel conditions/stability, and running normally with 7.0 kts. V/L position was stbd [starboard] beam of Pt. Maisí. Hurricane Joaquin just about 140 to 160 nautical miles NWxN from vessel's noon position."

As his ship closed in on the open waters of the Windward Passage, Gelera handed over the bridge to his chief mate, Henry Latigo, a broad-shouldered Filipino with a shaved head who, like Gelera, had been sailing professionally since he was a teenager. Normally after the day shift, Gelera would walk the ship, chat with the crew, see if there were any problems, check if anyone needed anything. He liked to joke with his men. "Now we're going too fast for you to catch fish," he might say to the deck crew. He'd also check on the engine room to see how things were running.

But today Gelera was tired, so he went down to the officers' mess to get something to eat. Lunch that day was turkey and plantains. Afterward, with a full belly, he headed to his cabin to catch some sleep. He said a little prayer as he lay down on his cot. At the foot of his bed was the black-and-silver cross he affixed to the wall of every ship he sailed on. Then he closed his eyes.

They weren't closed for long. The rolling and pitching were getting worse, but he was used to that. What disturbed his rest was a noise he heard. *Bang!* It seemed to come from either under the ship or somewhere on the side. It was loud enough that he was concerned. He got up and went down to the engine room to ask the engineer if he heard the sound. *No, Captain, didn't hear anything,* the engineer said, then added, *but it's loud down here.* Gelera climbed up to the bridge and asked Latigo if he'd heard anything. Nope. Gelera went back to his cabin. As he lay down, he heard it again. *Bang!* He sat up. Reflexively, he knelt down to pray again. *Lord, is this is a sign?* Then he got dressed and went up to the bridge.

Sailors are known for their superstitions. Whistling on board is an invitation to storm winds. Carrying bananas on a ship brings bad luck. (One theory holds that many ships lost at sea during the early days of the Caribbean trade were carrying bananas.) Personal grooming brings ill fortune. Mariner lore is filled with these beliefs. What they indicate is not necessarily that the fates are fickle, but that once at sea, a sailor can control only so much. So he tries to control the little things, in part because the things he can't control are so enormous: winds, waves, storms. As a result, his mind becomes attuned to the extrasensory for signs of what could go wrong—especially when he's the captain. Lives depend on it.

Gelera was not about to ignore this strange occurrence. Between his faith and his occupation, he was conditioned not to disregard the unexplained. In this case, he knew in his heart—he somehow just *knew*—that his ship was doomed. In that moment, he was convinced they would not reach port. He decided to act on his intuition, without trying to explain it to his crew.

Of course, it's possible that the banging sound the captain heard was very real—steel on an old ship giving way to waves, or cargo shifting violently—and the engineers didn't hear it because it was too noisy where they were.

Whatever had happened, Gelera wasn't taking chances. He went back up to the bridge. Surprised, Latigo asked why he was there.

"I don't know," Gelera said. "I cannot sleep." Then he told Latigo to order the crew to report to the bridge deck with their life vests.

Latigo, casting a quizzical look at the shipmaster, asked why.

"This ship will not reach its final destination," Gelera told him, a tone of certainty in his voice. "Call the crew."

This is reflected in the ship's log, which states that at 12:45 P.M., "Master advise c/off [chief officer] to advise all crews to s/by on the bridge with all survival gears." The ship was running fine, yet the captain was mustering his crew to their survival station. They must have thought that after all those years at sea, the man had finally lost his mind.

In retrospect, Latigo said he never thought his captain crazy. Perhaps overly cautious. But he's been following orders his whole life as a sailor, and he wasn't inclined to stop now. "He asked me about the noise, but I no hear it," Latigo said. "The captain have the power, he have the second eye."

The perplexed crew assembled on the bridge. Gelera told them that no one was allowed to go back down to their cabins, or to the galley. The exceptions were the two engineers who had to be in the engine room. However, Gelera still did not explain his reasons. They might not understand if he told them that he was responding to divine intervention.

The eight sailors assembled on the bridge. Some lay down on the deck outside and put their life preservers under their heads. Some napped. Some listened to music on headphones. Several complained that they wanted to go to their cabins and watch a DVD. The cook said he needed to get back to the galley to start preparing dinner. Again, the captain denied their requests.

It was about four o'clock when the ship passed Punta de Maisí and entered the Windward Passage. The weather immediately deteriorated. The wind grew stronger, the waves bigger, and the rain came down harder, obstructing visibility. As the ship put Cuba behind it, the crew was astonished at the sudden increase in wind speed. What the sailors didn't realize was that Cuba's rocky mountains had been providing leeward protection from Joaquín's winds, which, in their counterclockwise turn, were swooping down below Cuba, then curving up north again. As the ship headed into the open stretch of sea that was the Windward Passage, Joaquín's winds swept up from the south and came at their starboard beam ferociously. Big swells rocked the ship from side to side as the winds hit. But the ship also pitched, bow to stern. The wind's circular motion had created a confused sea, where the waves did not necessarily come from a uniform direction, compounded by strong currents in the passage. This is dangerous water to be in, because waves can join together and grow bigger.

The wind was blowing about 30 knots, Gelera estimated, with 50-knot gusts (58 miles per hour). Dark green-blue water rose up in malevolent lines rolling toward them like huge slabs of granite. The wind hissed across the crests of the waves, lifting the froth of the whitecaps off into spindrift. The ocean spray was so dense it was hard to see. Gelera estimates the ship was battling twenty-to-thirty-foot seas. Not all the waves were that high. But some were much higher.

As the waves rushed the ship, the captain steered the *Minouche* to take them head-on, hitting the throttle to zoom up the face. When the ship crested, he'd ease the throttle and slide down the back of the wave into the trough. As the ship descended, the horizon would disappear from view. As soon as the bow stabilized, he'd look for the next oncoming wave, altering the ship's direction to take each wall of water as directly—and therefore as safely—as possible.

"I told my crew we will escape this tragedy," Gelera recalled, "if you follow me."

He was convinced that the crew's ability to remain disciplined in the face of nature's fury would decide their fates.

They rode the weather like this for about an hour. Fighting the waves cut the ship's speed in half, to about four knots, slowing their ability to sail through the most exposed stretch of the pass and its dangers. Crew members were getting seasick. Cargo was knocked loose. The rocking back and forth had shifted the freight, causing the ship to list slightly to port. Gelera ordered the chief mate to get on deck, cut the lashings with a knife, and jettison some cargo in an attempt to correct the list. Latigo tried, but returned a few minutes later to report that it was too dangerous to go on deck. So instead, Gelera ordered the engineers to pump ballast into the starboard tanks.

Then, as the ship crested a big wave, Gelera looked down at his controls to see that the engine had stopped. The boat had lost propulsion. The captain suspected the wave had lifted the propeller out of the water long enough for it to spin freely, which would prompt an automatic shutoff. The ship rode down the huge wave without power.

The engineers scrambled to rouse the old engine, using manual overrides to try to coax it back to life. Nothing worked. Slowly the ship began to turn beam to sea. Gelera ordered the engineers topside. Then he called the ship's agents in Miami. He told them they were in an emergency: The ship had turned broadside to the waves, the port list was now about fifteen degrees, and, most critically, they had lost propulsion.

As he finished the call, Gelera looked out the window of the bridge and saw the wave. By now the thick gray clouds had choked out the sunlight. The ocean was a rough black slate, marbled by chaotic lines of whitecaps. Rushing toward Gelera and the *Minouche* was a dark mass of water that rose above the surrounding waves like a mountain in a range of foothills. As it neared, the wave began to rise and rise until it towered over the ship—Gelera thinks it was fifty feet high. Then it crested and came down squarely on their starboard side with a thundering crash. As the giant wave broke, its waters rushed over the decks in a deluge. The *Minouche* gave a convulsive shudder as a thousand tons of steel yielded to the overwhelming weight of the water. The force of the impact ripped out the metal plates securing the derrick's guy wires, or stabilizing cables, to the ship. The two cargo booms swung wildly over to the port side, yanking the ship down violently into a thirty-degree list. The waves continued the attack, relentlessly pounding the crippled ship.

Gelera and Latigo made a quick assessment. Without power,

they had only one option. The captain told everyone to put their life vests on and to muster at the poop deck. After securing his own life vest, he opened the covers on the emergency distress switches and hit them all: Inmarsat C, SSAS, EPIRB. He lit up the board.

Meanwhile, Latigo sprinted down to the captain's office and opened the safe. He grabbed the ship's money and the crew's passports and training certificates, and stashed it all in a plastic garbage bag that he gripped tightly. Then he ran back to the bridge.

The crew was miserable. Not only were they terrified as their ship slowly sank under them, but almost to a man they were seasick and vomiting, Latigo included. Only able seaman Jules Cadet and Gelera were not heaving. Gelera was too busy frantically taking care of last-minute tasks, including putting out a Mayday on the radio, to vomit.

It was 6:24 P.M. when Captain Ranjit Gokhale, on the *Falcon Arrow*, heard the Mayday call on channel 15. The signal was weak, but clear. Gokhale figured that meant the caller was on a handheld battery-operated radio. Gokhale responded.

Gelera communicated their situation—that the ship was listing, that some cargo had shifted on deck, and that the booms had come loose. Gokhale located the *Minouche* on the radar, noting that it was about twenty-four miles to the southeast. Without hesitating, Gokhale changed course and steered the *Falcon Arrow* toward the *Minouche*. He got on the radio and asked Gelera how many people were on board and if anyone was injured. Then he relayed the information to the U.S. Coast Guard's District 7 Command Center, which had also received the *Minouche*'s distress signals.

As the *Falcon Arrow* sailed down from the northwest, another ship, the *Cronus Leader*, a seven-year-old 652-foot Panamanian-flagged ship, also heard the Mayday and responded. It was coming from the south about fifteen miles away. Gelera warned both ships that there was cargo in the water and they should be careful. Gokhale eased his ship's speed and ordered members of his crew to scan the sea for debris. As soon as the *Falcon Arrow* slowed down, reducing its ability to slice through the rough ocean, twenty- and thirty-foot waves started crashing against the hull, sending spray over the deck. The sun was going down now and the clouds hastened the darkness. But as they neared their target, Gokhale could see the emergency lights of the *Minouche* glowing in the waves.

When his ship was about two miles away, Gokhale examined his options, which, from the bridge of a towering six-hundred-foot ship, were limited. The *Falcon Arrow* was enormous and difficult to maneuver. If he sailed any closer, there was a risk that the waves would cause the two ships to collide, and the water was too violent to launch a smaller boat to try to rescue the men. Instead, he ordered the *Falcon Arrow*'s engines cut to near idle and sailed into position on the *Minouche*'s starboard side, to act as a wind- and wave break.

The captain of the *Cronus Leader* came to the same conclusion and ordered his ship to hold its position about a mile away. The most either ship could do at this point was to simply keep watch until help arrived. Gokhale told Gelera he'd stay on the radio with him the whole time, and asked Gelera to let him know when they were getting ready to abandon ship.

On the *Minouche,* the crew braced themselves against the rails, trying not to slip on the soaked deck. They could see the lights of the big ships twinkling in the rain and waves. It was a comfort, no doubt, that they were being watched—that there were other sailors nearby who knew of their plight and were concerned about them. But the presence of the big ships did little to dampen the growing terror inside them. The ships could not reach them, and they knew that soon they would be in the water, in just a life raft, at night, getting hammered by a hurricane.

From the deck of the *Minouche,* Gelera fired flares that opened in the night sky like fireworks in a celebration. They were anything but. *Here we are,* their incandescence declared. *Help us if you can. If you can't, mark this spot as our grave.*

The crew assembled around the starboard life raft, which was contained in a barrel the size of an oil drum. When released into the water, it would automatically inflate into a floating tent that would provide some cover from the waves. Ever since the engine stopped, the waves had been turning the bow westward. Gelera chose the starboard side of the ship because it was now protected from the winds. But when the captain launched the two-hundred-pound raft, everyone watched in horror as it hit the water, exploded into form, and was picked up by the wind like a child's kite and blown out to sea. It remained tethered to the ship, but was now flailing wildly in the waves. The men couldn't reach it, and it was too unwieldy to try to pull it back to them.

An instant of panic flitted across the captain's mind as he watched the winds toy with the raft. But he recovered just as quickly, and started shouting directions in his high-pitched voice, ordering everyone down the listing deck, where there was another raft. The men crabbed their way across the slippery metal. And then, when they reached the port rails . . . they had to wait. The waves had turned the ship so that the wind was hammering the port side. If they launched the life raft here, it would blow into the sinking ship and the sailors wouldn't be able to push it free. Fortunately, the waves, in their chaos, were now turning the ship so that the wind would soon be on their starboard side. They just had to be patient.

It took all the resolve and discipline a sailor could muster to remain calm as those minutes passed. From above, the rain thrashed them while the swells breaking against the side of the ship erupted in curtains of spray. From below, the waves seemed to beckon, a thousand white-capped arms motioning for the sailors to join them. They tasted the salt from the spray, washed away immediately by the driving freshwater of the rain.

Finally, after about thirty minutes, the *Minouche* had turned enough to launch the life raft. Gelera ordered the crew to stand in a straight line and grip the port rail. As the ship descended ever lower, Gelera finally gave the order. The drum dropped into the water and the life raft burst out of its container. This time it stayed near the ship. Now came the difficult part. Because the ship was rocking so violently, each crew member had to perfectly time his entrance into the raft, one by one. The men had to lock into the rhythm of the waves, a terrifying dance to a dire *konpa*, the Haitian meringue, played by the furies. They bent their knees as the raft dropped, and as it surged back up toward the ship, pushed off with their calf muscles and launched themselves in. One man at a time, they made it into the orange-canopied raft. Gelera, of course, went last. He held a knife in his hand, and as soon as he was safely aboard the life raft, he reached back, grabbed the tether, and cut it.

As the raft drifted away from the ship into the dark and vast sea, a thought too terrible began to creep into the sailors' minds: If thousands of pounds of steel couldn't protect them, what chance did they have with only a few millimeters of rubber and canvas separating them from the depths?

Chapter 9
"FLY SAFE"

By mid afternoon on Thursday, October 1, the storm had clogged the sky over Great Inagua Island and it was clear to the guardsmen about to start their 3:00 P.M. shift that no one would be going on patrol. Instead, rescue swimmer Ben Cournia, pilot Rick Post, flight mechanic Joshua Andrews, and some others piled into a pickup truck and went for a drive into Matthew Town to get a firsthand look at the growing tempest. Dense, roiling clouds the color of dread spat down needles of rain, and the wind bent the scrubby palm trees sideways. As the men drove toward town, the wind was blowing hard enough to lift shingles off the roofs of some of the houses. The truck passed the island's 113-foot lighthouse, where the waves were crashing with such ferocity the spray shot over the two-hundred-year-old tower and came cascading down the other side. The wave contact sounded like cannon shots. The men stopped the truck and stepped out to admire their perennial adversary, the hurricane.

Seven years ago that adversary won, when Hurricane Ike aimed straight for Great Inagua and forced the Coast Guard to flee. Ike was one of three back-to-back storms in the region that year, a juggernaut of natural violence. Clearwater's commanding officer, Richard Lorenzen, was the operations officer at the time. He recalled looking at the satellite maps and seeing the three storms lined up like bowling balls aiming straight for the Bahamas. The Coast Guard decided to initiate a massive evacuation of its two-helicopter base on Great Inagua and its one-helo base on Andros Island. This required seventy-two hours of continuous flying by a fleet of C-130s from Clearwater, a total of forty-eight sorties. Both island bases were completely emptied out—not just personnel and aircraft, but tow vehicles, dual-axle trailers, and entire fuel trucks, plus smaller equipment such as computers, tools, and weapons. The sky was darkening overhead on the third full day of the evacuation when a Coastie on the last sortie turned out the lights and locked the door to the base on Great Inagua.

Ike damaged nearly 80 percent of Great Inagua's homes but claimed no lives. As the storm peeled off the island, though, it gained strength, reaching Category 4 status, and went on a lethal rampage through Cuba, where it killed seven people. It then crossed into the Gulf of Mexico and killed twenty-one more in Texas, Arkansas, and Louisiana.

About two days after the storm passed, the Coast Guard was able to get a plane out to Great Inagua. The pictures sent back to Clearwater showed total devastation. The base was flattened. The hangar's metal doors and walls had been ripped off like tissue paper. Thick steel I beams were twisted and bent ninety degrees like twist ties.

But the base was too important to abandon, so the Coast Guard initiated plans to rebuild. Meanwhile, its people were temporarily relocated to Turks and Caicos. At first, the only place that could house and feed all the staff was a Club Med, but soon they found some cabins more befitting a taxpayer-funded agency.

This time, the base on Great Inagua would have to be sturdier. The service contracted with Oldcastle Precast, a company that specialized in prefabricated concrete structures, and in 2011 Oldcastle started building the twenty-thousand-square-foot aircraft hangar and hazardous-materials building so that it could withstand a Category 5 hurricane. The structure boasted four 28-by-31-foot concrete door panels, each of which weighed sixty thousand pounds and was mounted on a motorized track that guided them as they were opened and closed.

The men on Great Inagua would not be going out that afternoon as the hurricane's bands reached the island, but at least this time they wouldn't have to evacuate.

As Cournia stared out at the waves, he could feel the wind whipping the air around him and pushing him—not away from the water but toward it. These were the conditions he trained for. It was as if the storm's energy was charging the swimmer, activating something inside him. This, right here, is inspiration, he thought to himself. He glanced over at Post, then back at the enraged water. "Sir, I may sound crazy for saying this," he said quietly, "but I kinda want to go swimming right now." Post looked at him sideways. (On hearing this later, Commander Scott Phy, the operations officer in Clearwater, shook his head. "This is when the rescue swimmers need adult supervision," he said. "They always want to get in the water, no matter what.")

But no one was going anywhere anytime soon. By the start of their shift, the eye of Joaquin was about 120 miles away. They would shelter in place.

Once they were back on base, the four-man helicopter crew—Cournia, Post, Andrews, and Dave McCarthy—changed into their flight suits and reported for duty in the main hangar. They prepped their gear, including their helmets, gloves, kneepads, and search and rescue (SAR) Warrior vests. These are twelve-pound harnesses that hold most everything a guardsman would need to survive if he went down: an inflatable bladder, flashlight, signal mirror, flares, chemical light sticks, knife, personal locator beacon, radio, even rations—packets of water and chewable candies. Cournia's was slightly different. The rescue swimmer's vest, called a SAR Triton, carries about nine pounds of gear while also acting as the harness that attaches to the helicopter's hoist cable. It is a significant bit of weight when dropping into the water, but all rescue swimmers are required to wear it.

In the hangar, the maintenance crew topped off the Jayhawk's fuel tanks and did a preflight check. Then . . . everybody went to their concrete block bunkers. By now, anywhere outside was supremely inhospitable, so there was nothing to do but hunker down. The men went to the pilots' hooch to bake chocolate chip cookies and watch an episode of *Narcos* on Netflix. The day before had been Andrews's thirty-second birthday, and the men had been informed that the crew swap was canceled and they'd be on the island for the foreseeable future. The cookies would be a small comfort. Post, meanwhile, resigned himself to not making it back in time for his daughter's first birthday.

Complete relaxation, however, was out of the question. In the back of everyone's mind was the knowledge that, if there were any ships still out at sea, there was a very good chance they would need help. Word had already begun to spread about *El Faro,* a huge container ship that no one had heard from in a disturbingly long time.

When Captain Gelera hit the Inmarsat C distress button on the *Minouche,* it bounced off a satellite and landed in Falmouth, England, where it was promptly relayed to the U.S. Coast Guard's District 7 Command Center. In addition to the satellite message, the *Minouche*'s Miami agent, Richard Dubin, called the Coast Guard and forwarded the email he received from Gelera containing a description of the emergency. The Coast Guard also had the EPIRB data. In this instance, the EPIRB accurately broadcast the ship's location. The District 7 specialist on duty knew whom to call.

Back in Clearwater, the weather couldn't have been calmer as Commander Scott Phy walked to his Toyota 4Runner in the parking lot outside his office at the end of the day. The temperature was about 80, the sky clear and blue, with no wind to speak of. But all Phy could think about was the weather more than 760 miles away. The entire Seventh District was still on alert, waiting for some word from *El Faro*. The last message had been at seven that morning, and while the information communicated to district headquarters in Miami was that the ship's troubles were manageable, the storm had only grown worse. No one had heard anything since.

On the other hand, this wasn't particularly unusual. Big ships often lose communication in storms. There have been many cases where the Coast Guard sends a search plane out, locates the ship, determines that it's not in danger, and then drops some radios for the vessel to use in place of its downed communication system. It was rare to lose a big ship in a storm quickly, without any sign of trouble. Eight-hundred-foot ships are like floating cities. They don't just sink without a trace.

After dinner, Phy and his wife, Lisa, put their seven-year-old son and ten-year-old daughter to bed. It was a little before 8:00 P.M., and Phy was looking forward to sitting down with his wife and catching up over a glass of wine. At exactly that moment, the phone rang. It was the operations duty officer at Clearwater. They had a case. Phy quickly got the rundown: A ship at sea south of Great Inagua had sent a distress signal. Phy may have been expecting to hear about *El Faro*. Instead he heard about the *Minouche,* a two-hundred-plus-foot motor vessel with twelve crew, engine disabled, listing thirty degrees to port in heavy seas. The vessel was Bolivian flagged, nationality of crew unknown. Phy considered his assets. Clearly the Jayhawks on Great Inagua were within range, and District was giving the green light to Clearwater. *If you can get out there, go!*

The specialist at District 7 also relayed that the 270-foot Coast Guard cutter *Northland* was sailing to the site as fast as it could. But it was too far west to get there anytime soon. Additionally, there were two Good Samaritan ships at the scene, the *Cronus Leader* and the *Falcon Arrow.* Crew members on the *Cronus Leader* reported witnessing the *Minouche*'s crew inflate a life raft and abandon ship. But the seas were too violent to launch smaller boats for a rescue, and the waves were too rough to get close to the desperate sailors. The crashing waves could smash the tiny life raft, and everyone in it, into the side of the larger vessel.

Phy kissed his wife, told her he'd take a raincheck on the wine, and cleared off the living room coffee table. He had the TV on to monitor the storm. Then he set up his laptop and loaded the relevant weather imagery and data. He monitored another weather site on his iPad. Next to him, he opened a green notebook, where he started logging updates.

A few minutes after orienting himself, Phy called Great Inagua and asked to speak to Dave McCarthy, the station's ranking officer, who also happened to be on shift to fly that night. "Lieutenant," he asked, "what's the visibility and cloud deck?"

"Visibility was about one to two miles in daylight," McCarthy answered. "And before sunset, the cloud deck had been about three hundred feet."

The big question Phy had was, Could they fly out in this weather?

It may be a maritime service, but flying is part of the Coast Guard's DNA. Surfmen from the Kill Devil Hills Life Saving Station in North Carolina helped Wilbur and Orville Wright carry and launch their kite-like planes at Kitty Hawk for three years in a row, including the famed 1903 flight.

In 1916, at the dawn of aviation, President Woodrow Wilson created the Aerial Coastal Patrol. Coast Guard commander

Elmer Stone was the service's first aviator. He flew an aircraft called the NC-4, a bi-winged contraption with a skeletal frame and pontoons that looked like it was made of pipes and cloth. Nonetheless, Stone was part of a Navy team (the Coast Guard was under the direction of the Navy during World War I) that in 1918 attempted to fly a convoy of planes across the Atlantic. The Navy's NC-1, NC-2, and NC-3 didn't complete the journey, but Stone in his NC-4 made the trip in nineteen days, stopping along the way in, among other places, Nova Scotia and the Azores. He landed in Lisbon, Portugal, on May 27, 1918. Two weeks later, his achievement was overshadowed when two Royal Air Force pilots completed a nonstop trans-Atlantic flight in the other direction. Still, Stone was able to claim his place in the record books, establishing an air-speed record for amphibious aircraft and, later in life, helping conceptualize and develop the catapult launch system that enables planes to take off from aircraft carriers.

Stone became a proponent of using aircraft for search and rescue missions in the postwar years, which were lean, prompting some critics to wonder if planes were simply an expensive new toy the government could ill afford. In 1920, a Coast Guard air station in North Carolina had to close due to lack of funding. Gradually, though, the Coast Guard proved its worth and the government began relying on it more and more—anti-rum-running patrols during Prohibition, coastal anti-submarine patrols in World War II, drug patrols in the modern era, and the ever-present search and rescue missions.

Like its early embrace of fixed-wing craft, the Coast Guard was quick to see the potential of the helicopter, a much more useful and nimble tool for search and rescue missions than a plane, and was instrumental in the aircraft's development. As soon as Coast Guard brass saw the public flight of Russian-born Igor Sikorsky's prototype in 1940, they put in an order. In 1944, the Coast Guard conducted the first helicopter landings on a

ship. The service's first helicopter pilot, Frank Erickson, conducted the first helo rescue mission later that same year. Erickson
and the Coast Guard's helicopter detachment were stationed at
the Sikorsky Aircraft plant in Connecticut to get trained by and
work with the famed helicopter designer. Erickson collaborated
with Sikorsky to develop power hoists for helicopter rescue missions and pioneered the idea of using stretchers to evacuate the
injured.

Helicopters and their crews grew better, stronger, and faster
over the ensuing decades, which is why McCarthy assured Commander Phy that getting the Jayhawk up and out was doable. In
a worst-case scenario, McCarthy said, they could fly out, assess
the situation, and return if they needed to. Phy explained that
they were looking at a rescue from a life raft, possibly deploying
a rescue swimmer. The men discussed how much fuel they would
need for twelve hoists out of the water, and where to take the
rescued sailors. The MH-60 Jayhawk has a seven-hundred-mile
range, plenty of fuel for the distance they were talking about,
even in conditions where they would be burning fuel simply to
fight the weather and stay in place. One option was to hoist the
survivors onto the helicopter and then lower the men onto one
of the two Good Samaritan vessels nearby. The *Falcon Arrow*
reported that it had room for a helicopter to land on its deck. The
Coast Guard ship *Northland*, meanwhile, was headed to the
scene and also had a helicopter landing pad on it; that might be
an option as well. Also under consideration was a plan to hoist
the survivors onto the helicopter and fly to the nearest and most
sensible landing point. That might be Great Inagua, or it could
be Guantánamo Bay, Cuba. At that point they would have to
determine what to do with a crew of unknown nationals, but
that was the least of their concerns. "Regardless of the crew's
nationality," Phy told McCarthy, "just get them to the safest
place you can."

The commander then told his lieutenant to stand by while he

called the skipper. He hung up and dialed Captain Lorenzen's cellphone. Phy needed permission to "prosecute the case," as this stage in mission planning is called. Lorenzen was home and had just finished dinner. He stepped outside and sat down on his front steps to take the call. Lorenzen, a Jayhawk pilot himself, listened carefully as Phy laid out the variables. This was not the captain's most seasoned crew, but he knew them all. He had flown with both pilots, Dave McCarthy and Rick Post, and had done a mission with Cournia. They were all highly competent. His only hesitations were that Post was fairly new and had never seen this kind of weather, and McCarthy was in the Jayhawk after spending a couple of years flying a different aircraft. Still, they didn't put a crew together if the brass didn't think they could get the job done. "When we say 'Go,' we have full confidence they are ready to go," Lorenzen recalled.

As for the aircraft itself, Lorenzen was supremely confident that the Jayhawk could carry the men and any survivors safely in this weather. But he wanted regular updates on the storm. The days of "You have to go out, but you don't have to come back" are long gone. Once he was satisfied that he wasn't sending his men on a suicide mission, he gave the okay.

Phy was a C-130 pilot, not as familiar with the world of helicopters. He was nervous, but he, too, trusted the men and the machines. Phy rang McCarthy back. "Okay, go out and fly safe," he said. Then he went into the kitchen to make some coffee. It was going to be a long night.

His orders in hand, McCarthy hung up the phone. He had no doubt they could get the Jayhawk out there. The helos were tough and this is what the men were trained to do: fly in extreme conditions. And the *Minouche* was only about forty miles away. Still, conditions were deteriorating. Joaquin was holding steady as a Category 4 hurricane, and it had continued moving south toward Crooked Island. The eye was now about one hundred miles north of Great Inagua. Tropical storm–force winds easily stretched over the rescue site.

McCarthy mustered the helicopter and ground crews in the big hangar for a briefing. He explained the mission: cargo ship down, forty miles southeast, crew abandoning ship. He ran down a list of known risks. The main one, of course, was the weather— driving rain, high winds, extensive cloud cover. Then he asked if anyone had reservations. Silence.

They may have assessed themselves as ready, but the truth was, this was virgin territory for the helicopter crew. None had flown in conditions this extreme before, except maybe Cournia during his tour in Alaska, much less conducted a search and rescue operation in a hurricane—at night. Most civilian helicopters won't take off in 20-knot winds. The MH-60 has a wind limitation of 60 knots off the nose and 45 from any other direction for takeoff. The Coasties were heading out in 35-to-40-knot winds with gusts up to 60.

First, there would be a slight assignment shift. For two weeks now, McCarthy had been flying patrols with Rick Post. He knew his copilot's skill level and trusted him. McCarthy decided to have Post pilot the aircraft. This meant that Post would get them to the scene and then communicate with Andrews, the flight mechanic, to establish the right positioning during the hoisting operation. This would give McCarthy greater situational awareness and allow him to oversee the whole operation. With Post at the controls, McCarthy would be freed up to monitor the fuel levels and the weight inside the helicopter, and maintain communications with the base, survivors, and the Good Samaritan ships.

McCarthy and Post then discussed the most immediate problem. Visibility would be near zero, leaving the pilots vulnerable to vertigo or spatial disorientation. Coast Guard pilots train for these conditions in simulators, so they were somewhat familiar with the effects and what the onset might feel like. They discussed how to approach the problem and decided to wear their night vision goggles immediately upon takeoff. This would limit their peripheral vision, but they would be able to get at least a faint glimpse of the horizon, which would help them orient themselves. Meanwhile, the other crew members were focusing on tasks specific to their jobs.

"Abandoning ship." Joshua Andrews let the words sink in. Things were about to get very real. As much as you train and prepare for this moment, when it actually arrives, you can't help but question if you're ready. Andrews told himself to slow down, take a deep breath, and focus on the job. *You've trained a thousand times for this.*

It helped that battling weather was in Andrews's blood. His father was a "storm trooper," an insurance adjuster in Boerne, Texas, who specialized in areas damaged by natural calamity. When other kids were playing baseball in the summer, Andrews was climbing roofs with his dad to measure holes made by tornadoes. So, in a sense, this work was his legacy.

Andrews jogged over to the Jayhawk for a quick assessment. He checked the hydraulics and oil levels. He examined the cable on the hoist. He looked inside the cabin for ways to maximize space. In a best-case scenario, they would be flying back with twelve survivors. First thing to go was the big water pump, used to bail out sinking ships. This ship was already going down, he reasoned. Removing the pump would save weight and create space. Then he packed dozens of extra chemical lights. They would not only be fighting the wind, rain, and seas, but also the dark. Post, meanwhile, followed behind, acting as a second set of eyes, making sure the Jayhawk was fully kitted out and ready to go.

Cournia changed into his three-millimeter neoprene "shorty" in the hangar, a bright orange wetsuit with short sleeves and legs. Air temperature was in the low 80s, the water in the mid-to-high

80s, so he wasn't worried about getting cold. He strapped on his SAR Triton vest. Then he grabbed his swimmer's helmet, his fins, mask, and snorkel.

Still inside the hangar, the men climbed into the Jayhawk and strapped into the five-point harness built into each seat. Outside the wind was screaming around the concrete walls, and the rain was ferociously lashing the roof. The ground crew attached a tow bar to the tail of the Jayhawk and fired up the MT3 tow vehicle, affectionately called a mule. Then the massive hangar doors, those sixty thousand pounds of standing concrete built to withstand Category 5 winds, winched open on the motorized chain. Outside, the lights on the tarmac flashed intermittently through the rain as the mule towed the Jayhawk to the launch apron. Once there, the ground crew rushed into the wind and downpour to uncouple the mule and drive it back into the hangar.

In the cabin of the Jayhawk, Post toggled the ignition switches and the key turned. The fifty-three-foot rotors began to spin. The two pilots slid their night vision goggles down, cutting off their peripheral vision and suffusing the world in front of them in a dull green illumination. Then they initiated liftoff into the near-total darkness of a howling storm. It was 9:27 P.M.

Helicopters fly in three dimensions—up and down, left and right, forward and backward—which means that piloting an MH-60 Jayhawk requires all four limbs. The pilot's left hand operates the "collective," a lever that controls the pitch angle of the rotors and the up-and-down lift of the aircraft. Both feet work pedals that account for the "yaw," or anti-torquing motion, by controlling the pitch angle of the tail rotors. And the right hand maneuvers the "cyclic," a joystick between the knees that manages pitch and roll. There are three screens below the windshield in front of each pilot and a cockpit bristling with buttons. In a very real sense, a Jayhawk pilot *becomes* the machine.

As Post lifted the helo above Great Inagua, the black night, compounded by the storm, enveloped the men. Post followed the lights along the coastline south then east. The radio guard at the Coast Guard base in Nassau crackled on. Every fifteen minutes, the Jayhawk pilots were required to report their position. Meanwhile, the two Good Samaritan ships with eyes on the *Minouche*'s life raft were on the radio, keeping everyone apprised of events at the scene.

From the back of the Jayhawk, Cournia and Andrews watched the pinpoint lights of Great Inagua quickly recede behind the curtain of rain until they were staring into an impenetrable black void outside. Soon there was nothing but the sounds surrounding them: the thud of the rotors, the crackle of the radios, and the howl of the wind, as sudden gusts pushed them forward or up, then dropped them down or yanked them from side to side. Again, training was critical here. The men focused on the jobs in front of them, and, for the most part, ignored the fact that at that point they were nothing more than a chew toy in the jaws of nature.

Chapter 10
BATTLE RHYTHM

The life raft was a twelve-man SOLAS Class A hexagon, orange on top, black on bottom, with a zip-up enclosure to keep the water out and a blinking strobe light mounted on top. Inside, the crew of the *Minouche* cowered in fear and nausea. Everyone, it seemed, was seasick. Vomit sloshed with seawater on the raft's floor in a vile soup. Men were moaning. The mood was deteriorating, and with it the men's will. Gelera knew he had to say something. *Okay, get sick,* he said, *but no shouting, no panicking.* He said this crisply, as if he were still on the bridge. *If you do, we will have a problem.* Then he tied the EPIRB and the SAR transponder to lines inside the life raft.

Waves crashed down on them. The wind pounded and roared from outside, as if desperate to get inside. The worst were the big waves that would break over them, pushing the raft underwater for a few moments before its buoyancy popped it back up to the

surface. Inside the raft, the concussive force of these waves
snapped the sailors' necks back and forth. Then Gelera realized
he could hear the swells as they approached, like boulders rolling
downhill toward them. He devised a system where he'd unzip a
ventilation flap a little bit, put his ear to the opening, and time
the impact. When a wave got close enough, he would tell the
crew to duck down. The wave would crash down on them, driving them below the surface. And each time, so far at least, they'd
popped back up.

Able seaman Jules Cadet thought for sure the waves would
shred the life raft and they would all drown. Cadet's eight children waited for him back in Haiti, and he wondered how his wife
would take care of their brood without him.

It was already past sunset. Joaquin had made sure that any
lingering light was completely blotted out. The crew bobbed at
the mercy of the waves in almost total darkness. They had flash-
lights in the raft, but they were concerned about saving the bat-
teries, so they used them sparingly. Strobe lights blinked on their
life vests. Every minute contained a conscious or unconscious
calculation of mortality. Would they be rescued? Would the life
raft break apart in the waves? Would the waves bring a floating
piece of metal down on them from above?

Many of the men, in a very Haitian way, were already accepting their fate as God's will, and were asking their mates to deliver
messages of love to their families if they happened to survive.
Gelera, whose religious devotion would be hard to match, also
trusted that whatever happened would be God's will. But he
couldn't resign himself to doom for a simple reason: He was responsible for everyone on his ship. It was his duty as captain to
keep them alive, and it would be his legacy if they perished. This,
perhaps, was an even greater motivating force than his own survival. A captain who lost his crew would leave a stained memory
on this earth.

Luckily, the old shipside hierarchies were holding. Amid their fear, the men did something that came instinctually to veteran sailors: They followed orders.

Then, through the wind and crashing waves, they heard an unmistakable thud-thud thudding sound.

The Jayhawk faced a headwind, so it took about forty minutes to reach the *Minouche*'s coordinates. It was after 10:00 P.M. as the helicopter closed in. Below, the crew could see the outline of the freighter awash in waves. The deck and navigation lights were on, and the ship, listing on its side like a wounded beast, shimmered ethereally in the dark. The helicopter swept by in a wide circle, scanning the surface for the life raft as well as floating cargo or other hazards. When they came back around, the ship's decks were now underwater, glowing a dull amber. The Jayhawk took another orbit. This time, on their return circuit the ship was gone, only its dimming lights visible beneath the waves.

The speed of the sinking shocked the chopper's crew. This was a big ship. Humans relied on big ships to stay on top of the water, to bring them their coffee and computers, or their rice and an old mattress. Now it was gone. In the vastness of the sea, all that size and metal had been scant protection against the storm.

As Cournia stared at the ghostly glow of the fading deck lights below, whatever enthusiasm he'd had for jumping into the stormy water drained from him. His job at that point was to use the helicopter's electro-optical infrared camera, or EOIR, to search for the life raft. Eventually he spotted a blinking light off in the distance. His pulse quickened. "Target located," he announced on the Jayhawk's internal radio, which the crew used to talk to one another. He was pointing the camera about a mile off in the distance, where a distinct shape bobbed in the waves.

Post flew the Jayhawk over, but rain squalls and waves blocked the raft from view. "Target lost," Post called out, and Cournia again scanned the dark water. This pattern repeated a few times as Cournia and Post would locate and then lose sight of the raft. Meanwhile, they established radio contact with Gelera. "Our swimmer will be down soon for you," McCarthy told the captain. They were not out of the water yet, but Gelera allowed a little wave of relief to pulse through him.

Eventually, Post was able to position the aircraft over the blinking light on top of the raft's canopy as it bounced wildly in the waves. Post tried to steady the Jayhawk into a hover. This was difficult. The wind kept pushing the helo around, and the dark messed with his equilibrium. Sheets of rain blinded the pilot. Instead of looking out the windshield to get his bearings, Post had to rely on the Jayhawk's hover bars on the instrument panel, which showed the helicopter's relation to a fixed point. Andrews tossed some flares out of the aircraft to mark the raft's location, but the flares drifted away too fast. When they turned on the main searchlight and beamed it downward, the light reflected off the rain back into the cabin, further blinding Post. They had to rely on a secondary light mounted on the Jayhawk's belly.

Once they had a firm handle on the raft's location and were confident they weren't going to lose sight of it again, the crew discussed the best approach to carry out the rescue. In these conditions, it was clear that Cournia would be deployed. The water was far too rough for the survivors to load themselves into the rescue basket. The question was whether or not the swimmer would detach from the hoist cable. In one scenario, he could descend on the cable, grab a survivor, and both would be hoisted up together. In another scenario, Cournia would descend, detach, and send the cable back up, where a rescue basket would be attached and lowered. He would then load the survivors in, one at a time, to be hoisted into the Jayhawk.

McCarthy, meanwhile, was trying to figure out what to do with the survivors once they were extracted. He called over to the *Cronus Leader* and *Falcon Arrow,* checking to see if they could launch rescue vessels. Both said the seas were still too rough. He asked if they had space for the helicopter to land on deck if necessary, or at least lower survivors down on the cable. Gokhale on the *Falcon Arrow* reiterated that he had the space, about one hundred meters of open deck.

Andrews didn't like any of these options and spoke up. Depositing survivors on a ship in these seas was too dangerous. McCarthy agreed. Ultimately the crew decided to lower Cournia into the water, have him detach from the cable, and swim to the life raft. Andrews would recover the cable, attach the rescue basket, and lower it. Then Cournia would take the survivors out one at a time and load them into the basket. Once the survivors were safely in the Jayhawk, Post would fly them back to Great Inagua.

McCarthy radioed Gelera on his handheld VHF radio and advised him of the plan.

Inside the life raft, Gelera translated what the Coast Guard told him into Haitian Creole. The crew was getting panicky. Someone shouted out *I have to go first. I'm going to die!*

"No one is going to die," Gelera replied calmly.

On the helicopter, Cournia took his flight helmet off and put on his yellow swimmer's helmet. He strapped on his mask and fins, and clambered to the door. Andrews handed him the hook on the end of the cable, and Cournia clipped it to his harness. Then Andrews opened the helicopter's door. A blast of wind rushed in. Cournia looked down at the waves thirty feet below him. His stomach dropped.

But whatever gripped him momentarily, fear or the instinct for self-preservation, there was never any doubt in his mind that he would jump into the water and do his job. He had trained hard for the right to be where he was at that moment.

Ben Cournia was thirty-six years old. He had been a member of the Coast Guard for fourteen years and a rescue swimmer for the past eleven. It was a very physical job with a built-in expiration date. Cournia didn't know exactly when he'd get too old for it, but he knew he was closer to the end than the beginning.

He had grown up a self-described "water rat" in the northern Minnesota town of Bemidji. His family lived on a lake, and he'd learned to swim before he could walk. He joined the high school swim team and ran cross-country. Endurance sports were his thing. He liked pushing his body to the limit.

A documentary about the Coast Guard on the Discovery Channel inspired him. He joined on January 3, 2001, a young man looking to escape a small town. After boot camp, he served as a seaman on a buoy tender in Alaska's Prince William Sound. Needless to say, between the job and the frigid water, he wasn't getting a lot of swimming in. When he inquired about the service's rescue swimmer program, which had always been his goal, he got some bad advice. An officer told him that because he didn't have twenty-twenty vision, he was ineligible. "I kind of talked to the wrong person," he said.

A year later, he was participating in cutter surface swimmer school. These are swimmers who ride on ships and are trained to aid people in man-overboard or search and rescue operations. It was a four-day class, and Coast Guard rescue swimmers, the service's elite operatives, were conducting the training. "When they saw me swimming, they were like 'What are you doing here?'" he recalled quietly, relating the anecdote as a fact more than a boast. "I was swimming circles around everyone. That's when I found out my vision didn't matter." He could just wear contacts.

He had to take an aptitude test and was put on a two-year waiting list. Finally, about three years after joining the Coast Guard, he was sent to Elizabeth City, North Carolina, to the rescue swimmer school at the Coast Guard's Aviation Technical Training Center.

The assignment was a professional triumph, but the timing wasn't ideal from a personal standpoint; there was a girl. Cournia had met Lindsay through mutual friends while she was attending Bemidji State University in Cournia's hometown. They were friends at first, painting the walls of their church together during a community service event and attending a music festival. He was handsome and more interesting than the other boys her age. She liked that he was adventurous enough to leave Minnesota. And he could tell a story, sometimes a fish story. She still remembers the one about the time up in Alaska when he caught a halibut as big as he was. Their blossoming friendship turned out to be a good thing when he shipped off to rescue swimmer school. It gave him someone to confide in after days of brutal training.

He invited her to his graduation, which, she recalled, was a bit awkward because she wasn't really his girlfriend yet. The next year, when she was a junior in college, he flew home around Christmastime. He visited her on campus and, to get some privacy from her roommates, asked her to take a walk with him—in a blizzard. "That's Ben—he has no game," she said. "That's what I love about him."

They dated for about a year, during which they saw each other in person for only about four weeks total, and got married in 2005. Their first home as newlyweds was in Kodiak, Alaska, where Cournia so regularly caught cases that his supervisor, John Hall, started calling him "God's helper." There was the pregnant woman evacuated off a fishing boat after getting mauled in a fish-processing machine while cleaning it; there were the survivors he rescued from a downed floatplane.

Cournia's training at the rescue swimmer school was the most rigorous program in the Coast Guard. Aviation survival technicians (their official name) are the branch's version of the Army Rangers or the Navy SEALs—except instead of learning how to kill with stealth, they learn how to save people.

The U.S. Coast Guard's rescue swimmer program was born from one of the worst shipwrecks of the modern era. On February 9, 1983, a cargo ship named the SS *Marine Electric* left Norfolk, Virginia, with a hold full of coal destined for a power plant in Somerset, Massachusetts. As it steamed north, it encountered a powerful winter storm. Forty-foot seas rolled the ship, submerging the bow and flooding through rusted hatch covers into the holds. The *Marine Electric* could not recover. On the evening of February 11, the captain ordered the crew to abandon ship. But before they could scramble into the lifeboats, the ship capsized. Many clung to life rings or other floating objects in the frigid waters. One crew member recounted swallowing "gallons of oil" that had spilled into the water.

At 4:00 A.M. on February 12, a Coast Guard HH-3F Pelican helicopter lifted off from Elizabeth City. At the scene, the helicopter crew lowered a rescue basket to the sailors treading water. But by that time, the horrified helicopter pilot realized, sailors were too numb to pull themselves into the basket. The Coast Guard had earlier radioed for help from the Navy, which had a rescue swimmer program. The Navy scrambled a helicopter, and it got to the scene a little after 6:00 A.M. Rescue swimmer James McCann descended into twenty- to forty-foot seas; the temperatures were so cold the seawater froze on his mask. He swam for hours, trying to help as many as he could. It wasn't many. Of the thirty-four crew aboard the *Marine Electric*, thirty-one died.

The country was appalled and demanded answers. Congress held special hearings to, in the words of an official Coast Guard history of the event, "question why the world's premier maritime rescue service was unable to assist people in the water." Shortly after that, Congress mandated that the admiral of the Coast Guard "establish a helicopter rescue swimmer program for the purpose of training selected Coast Guard personnel in rescue swimming skills."

The first Coasties in the program were trained alongside Navy candidates by the Navy's rescue swimmers. In 1988, a Navy recruit drowned during a training exercise. After that, the Coast Guard initiated its own training program in Elizabeth City, as part of the Coast Guard's Aviation Technical Training Center. They chose the motto "So others may live."

But before the would-be rescue swimmers can call themselves U.S. Coast Guard aviation survival technicians, they must first survive the training school. It's not uncommon for two-thirds or more of a class to drop out. Officially, the Coast Guard states that the attrition rate is 50 percent. Anecdotally, it is often much higher. Cournia, for instance, entered with a class of nine. Three graduated, one of whom had started in another class of twelve. When all eleven of his original classmates dropped out, this one remaining recruit transferred to Cournia's class. A recent class of sixteen graduated four swimmers; in another, only five of twenty-four candidates made it through the process.

The physical conditioning is relentless. Cournia recalled waking up at 5:00 A.M., eating only a small breakfast ("otherwise you'd throw up"), then running, running, running: sprints, twelve-mile runs, running through waist-high water. Then lunch, and swimming, swimming, swimming: 500-yard sprints, 500 yards underwater (in 25-yard intervals), 3,000 yards of laps. Add to this the survival technique classes, which the recruit cannot fail if he has any hope of graduating. In these classes, swimmers are confronted with simulations—a downed pilot wrapped in a parachute, a crew from a sunken fishing boat—and must figure out the rescue on the fly. There are "water confidence drills" in which an instructor swims up to you and basically tries to drown you. The school boasts fans that can replicate the winds in a Category 1 hurricane, a wave-making machine, and speakers that can blare thunderclaps at tremendous volume.

As relentlessly exhausting as the physical training is, the mental tests might be worse. Instructors repeatedly try to undermine the swimmer's confidence, constantly encouraging him to quit, telling him he's not good enough.

"Your body can adapt physically," explained Senior Chief Scott Rady, who coleads the school today with Cournia's former supervisor John Hall. "But mentally, that's the one you have to overcome."

In fact, most of the candidates who don't make it through the program "self-select out," as Rady put it. Another percentage can't continue because of physical injuries. Many of these candidates do return and eventually graduate. (This may be the reason the Coast Guard cites a 50 percent attrition rate.) Since the program started almost thirty years ago, a total of 940 aviation survival technicians have made it through. Today there are 360 rescue swimmers spread out among twenty-six Coast Guard air stations. Three of them are women.

Now, as Cournia peered down at the roiling sea from the open door of the Jayhawk, he was relying on every shred of his training and conditioning to override his body's normal response to danger. He had to go where it was not logical to go. He sat on the edge of the metal deck, his legs out the door, his eyes fixed on the life raft bobbing in the waves below. He reached up and snugged his mask to his face, then clamped his teeth around the snorkel's mouthpiece. Andrews checked the cable clipped onto the swimmer's harness.

"All right, swimmer's at the cabin door," Andrews reported into the cabin's radio. "Ready for harness deployment of the rescue swimmer."

"Roger, you may begin hoist," Post responded.

From his seated position, Cournia gave the thumbs-up and then pushed himself out the door into the air.

"Swimmer's going out the cabin door." While the pilots struggled to hold the helicopter in a hover, Andrews would be their eyes and ears for what was happening below.

As Cournia descended on the cable, he twisted in the wind. The water beneath him heaved back and forth. Big, long swells surged and then relaxed. Steeper waves crested and broke, causing the surface to erupt in a foamy effervescence.

"Swimmer's on the way down," Andrews said. "Swimmer's in

the water. Swimmer's away. Swimmer's okay. Clear to move. Back and left thirty. Retrieving hoist."

"Roger," Post responded. "Back and left."

When Cournia hit the water the sea felt reassuringly warm, but the strength of the swells and the ferocity of the waves caught him off guard. The water lifted him up and slapped him in the face. Waves crashing over him kept flooding the snorkel, forcing him to blow lungfuls of air through the tube to clear it. Still, he was in his element, and his apprehensions lifted as he steadied himself against the swells. He lifted his head and spotted the life raft a few yards away. Arm over arm, he dragged himself through the water in a freestyle crawl, swimming over and under the waves until he reached it.

Inside the orange floating hexagon, Gelera remembered what amounted to someone knocking on the door, a tap-tap on the curtain of the enclosure. He unzipped the curtain and saw a lanky guy in a swim mask hoist himself chest high into the raft, resting on his elbows. Cournia's exact words may have been lost to the adrenaline of the moment, but they were something like "Hi, I'm Petty Officer Cournia, U.S. Coast Guard. Does anybody speak English?"

The crew of the *Minouche* stared at him with wide eyes, tensed with overwhelming joy.

"Yes," Gelera answered, trying to contain his excitement. "I do."

"All right, sir, you're going to interpret for me. Is anyone hurt? No? Good. Okay, I'm going to take you out one at a time. Don't panic. I'm going to swim you over. The helicopter is going to drop down a basket. I'm going to put you in the basket. Make sure you keep your arms and legs inside the basket in a seated position."

Gelera repeated the instructions to the crew in Creole.

Something, apparently, was lost in translation. Gelera remembers the rescue swimmer saying something to the effect of, *Lis-*

ten carefully. I am here in order to save you! But I cannot guarantee that I can make it to all of you. To those who are going to survive, good luck. To those who are going to die, I'm very sorry.

"That's what I translated," Gelera recalled. "I understood him. At that moment, they [the Coast Guard crew] also put their life fifty-fifty." Gelera thought Cournia was telling them that it was a crapshoot if any of them would survive this ordeal.

(Needless to say, Cournia later recoiled at this. "I would never say that!" he said, then started laughing.)

Henry Latigo, the chief mate, asked Cournia what had happened to their boat. "It's gone," Cournia said, and was struck by the shocked expression that flitted across the men's faces. It was as if, bobbing alone out here in the storm, they were only now grasping the magnitude of their situation.

Cournia asked, "Okay, who's going first?" Gelera translated, and several hands shot up in the air. Gelera said, "I'm the captain; I'm not going yet. I want to go last."

Cournia pointed at the most scared-looking man, indicating he should come forward. The sailor did, and Cournia guided him down the boarding ladder of the raft and into the water. Then he grabbed the collar of the sailor's lifejacket, put the man in a cross-chest carry, and started swimming. Cournia could see the helicopter hovering overhead. He waved a chemical light in his hand to signal a pickup. As the Jayhawk swooped over them, the searchlight beam illuminated the cable and basket as if it were spotlighting a solo performance on a Broadway stage.

In the helicopter above, Andrews manned the hoist mounted outside the cabin door, with two hundred feet of stainless steel cable, made up of 133 strands of wire, wrapped around a motorized drum. Post may have been flying the bird, and McCarthy may have had operational control, but at this point, Andrews was giving the orders. He was the fulcrum between the pilots and the swimmer (and the survivors) in the water. He gave the directional commands so the pilots knew where to position the helicopter. And he was responsible for the swimmer's safety.

Down in the water, Cournia and the sailor waited for Andrews to lower the basket, which was just that: stainless steel bars welded together into a basket big enough for a grown man to sit in with his knees up. Post was still getting the hang of maneuvering the helicopter in the high winds and rain. As a result, the crew was having a hard time keeping the basket stationary on the surface of the water. Cournia would get close enough to reach for it, but then a big wave would sweep over them and carry the basket away. When the helicopter tried to move the basket closer, waves caught it and skipped it across the water. Cournia had to make sure that neither he nor the sailor got hit. Finally, after a few passes, he managed to seize and steady the basket. Then he pushed it down so he could float the sailor into it. Cournia reminded the sailor to keep his hands and feet inside at all times. He took a quick look underwater to make sure nothing was clinging to the basket, and finally signaled the okay sign to pull him up.

From above, Andrews was trying to keep his eye on how the swimmer was functioning in the water, which was difficult because it was dark. Andrews could see only the reflective tape on the raft and on Cournia's gear, along with the chemical light attached to his black mask and the strobe light flickering on top of the raft. Protocol was for someone's eyes to always be on the swimmer. So when Andrews had to turn his attention to unloading the first survivor, he asked the pilots, "Eyes on swimmer?" They answered in the affirmative and Andrews grappled with unloading the Haitian sailor. Once the man was out of the basket, Andrews pushed it back out the helicopter door and lowered it. After it hit the water, he saw a strobe flashing in the dark. This was the swimmer's emergency signal for a pickup. Andrews relayed this to Post and McCarthy immediately.

"Initiating emergency pickup," Andrews said.

"Roger," Post replied.

When Andrews looked back down, he saw the strobe sinking in the water, flashing more and more faintly. For an instant, he

felt a surge of horror. But then he saw Cournia, who had just
finished loading the second sailor into the basket and was swimming back to the raft. With a sigh of relief, Andrews realized the
light must have been a strobe that had fallen off the survivor's life
vest.

Andrews reeled in the cable and pulled the second sailor into
the cabin, still in the basket. Once inside, though, the sailor,
barefoot in cutoff jeans and a T-shirt, cramped up with fear. His
eyes were wide and he refused to get out of the basket. Andrews
yelled at him to climb out. When that didn't work, he tried pry-
ing the man's fingers off the metal bars. Finally, he unceremoniously flipped the basket over and the survivor tumbled out.

Andrews looked down for his rescue swimmer. Nothing.
Then he spotted him, furiously swimming after the drifting raft.

After sending up the basket for the second time, Cournia had
turned to swim back to the raft . . . only to find it wasn't there.
The wind and waves had pushed it more than one hundred yards
away. He cleared his snorkel, put his head down, and started
swimming. Waves crashed over him and pushed him down. He'd
surface and continue slashing the water in quick, decisive strokes.
After a while, he lifted his head, but the raft still seemed about a
football field away. He swam for a few minutes more, then
checked his progress. Again, the raft was barely any closer. It
seemed like it was drifting as fast as he could swim. Ultimately it
took about ten minutes of powering along in the crawl to reach
the raft, but he made it. He pulled out another sailor and signaled for a pickup. By the time this third sailor was loaded and
the basket was hoisted, the raft had drifted again. And Cournia
was getting tired.

None of this was lost on Andrews. He saw his swimmer constantly struggling to catch up to the raft, and he didn't think
Cournia would last the night if he kept having to give chase like

this. He consulted with McCarthy, and they agreed to pull the swimmer and regroup. Andrews sent the cable down and signaled for Cournia to hook in.

After they hoisted Cournia up, the crew debated different approaches. The solution they settled on was to "hover taxi" Cournia to the raft, meaning that the swimmer would dangle above the waves on the cable while Post flew him to the raft. Cournia would then detach and swim out the survivors. A relieved Cournia agreed that this sounded like the best plan.

The method worked well. Cournia was able to extract the next four survivors without incident. The dark cabin of the Jayhawk was filling with heaving bodies intermittently illuminated by the strobes on their life vests. They were dressed in shorts and T-shirts with no shoes. Most seemed dulled by shock.

It was now close to 2:00 A.M. They had been aloft for almost five hours. The crew was losing valuable time fighting the weather. And McCarthy was concerned about fuel. Down in the water, Cournia was worried about the same thing. Hoping to make up for lost time, he had begun speeding up his process, trying to get the sailors out of the raft faster and faster. When he went to the raft for the eighth sailor, even he had to admit he was going too fast.

Inside the raft, Gelera and the crew had calmed down as the rescue operation unfolded. Every time one of them would leave the raft, everyone else repositioned so their weight was evenly distributed. The raft was maintaining its position in the waves, but Gelera was worried that a big wave could still topple them over. When Cournia returned for the eighth time, the sailors could tell something was different. The rescue swimmer was much more brusque and direct. Cournia pointed at one man, a short but heavyset fellow who went by the nickname Maco. "You! Let's go!" Cournia shouted. Maco started to come forward, but then hesitated. Cournia gave him some space, but Maco refused to move. Maco, it turned out, didn't know how to swim. Gelera and the others were yelling at him in Creole to jump, *"Ale! Ale! Sote!"* But Maco couldn't bring himself to do it. He started to scoot back into the raft.

Cournia could hear himself getting louder and louder as his impatience showed through. He was thinking about the fuel. The Jayhawk carries about nine hundred gallons, which gives it a seven-hundred-mile range depending on the weather, or about six hours of flight time. But flying in this weather was burning more fuel than normal. Just trying to maintain a hover in the headwinds was taking its toll, and they needed to leave enough juice to get back.

Finally, fed up with Maco's hesitation, Cournia reached forward, grabbed the collar of his life vest, and pulled him into the water. The sailor exited the raft headfirst and splashed into the sea on top of the rescue swimmer. Maco gave a terrified yell and grabbed hold of Cournia, wrapping his legs and arms around him in a panicked clench. He pushed Cournia's shoulders down, struggling to stay above the water.

Cournia had a flashback to the training pool in Elizabeth City, battling an instructor during a water confidence drill. Now, with Maco, the events unfolded in a kind of slow motion that felt hyperreal. His training kicked in and he did what he had been conditioned to do: suck, tuck, and duck. He sucked in a deep breath, tucked his chin to his collarbone to protect his airway, and tried to duck out of the sailor's grip. The key was to stay calm. He wormed one arm free and gave Maco a gentle tap on his chest to let him know everything was okay. That didn't work. In fact it seemed to frighten Maco even more. The sailor thrashed about more chaotically, tightening his grip on Cournia, who was still underwater. So Cournia did what he was trained to do next, which was to take his free arm and jam his thumb into a nerve center under Maco's jawline. This allowed him to work his other hand free so that he could grab Maco's elbow and jam that thumb into another pressure point there. The sailor froze, and

Cournia used the moment to resurface and suck in a lungful of air. Back in control, he flipped Maco around in a front head hold. Then he signaled for the basket.

McCarthy, in the copilot seat, was monitoring all of the operation's moving parts. Each time a sailor was pulled into the cabin, he reported it to their radio guard, with whom the Jayhawk crew is expected to stay in contact throughout the mission. By now the cutter *Northland* was on scene and radioed in, awaiting instructions on how it could help. McCarthy was also monitoring wind speeds, the weight inside the cabin, the fuel-burn rate, and the radar. As time wore on, McCarthy realized the average time to recover each sailor was taking longer and longer, in spite of Cournia's efforts to speed things up. At this pace they wouldn't have enough fuel to stay aloft in order to recover all twelve sailors in one sortie. The Jayhawk was going to have to return to base and refuel. When McCarthy conferred with Post and Andrews, they agreed: It was time to head back. Andrews dropped the cable hook down to Cournia, signaling him to come up.

The swimmer did as he was told; he locked in and rode the cable up. But once inside the cabin, Cournia balked. "Why?" he asked. He was frustrated and pumped full of adrenaline. His job wasn't over and he wanted to stay with the raft. Andrews remembers him saying "There are still people there!" McCarthy understood where his swimmer's passion was coming from. He admired that. But as the officer in charge, McCarthy's first responsibility was for the safety of his crew. He denied the swimmer's request. McCarthy assured Cournia the sailors would be okay. The *Northland* was on the scene and able to keep close watch on the raft. This was an order. Post turned the helicopter around and pointed the nose toward Great Inagua.

Communication with the sailors on the raft was handed over to the *Northland*. However, when the guardsman on the ship tried radioing, he could hear only muffled static. The radio op-

erator gave instructions to Gelera, who held the radio, that if
they could hear him, to key the microphone twice. Two crackly
transmissions came through. The guardsman tried to sound as
calm and reassuring as he could. His job right now was to keep
the survivors from panicking. He explained that the helicopter
was low on fuel and had to return to base, but it would fly right
back. In the interim, the ship would monitor their safety.

The return flight to base took less than half an hour, shorter
than the flight out because they had a tailwind. They arrived
over Great Inagua without incident. As Post began the descent,
McCarthy thought he'd give his copilot a break, some time to
clear his head from the stress of maneuvering the Jayhawk in the
storm. So he took the controls for the landing. As they settled on
the tarmac and began taxiing to the hangar, McCarthy saw birds
huddling on the flight apron. Apparently they were weathering
out the storm there. Then, just as he was rolling toward the hangar, McCarthy saw one of the birds startle and lift its wings. He
stomped on the brakes. "No, no, don't do it!" the pilot yelled, to
no avail. The bird went straight up into the rotors.

It wasn't McCarthy's love of wildlife that had him screaming
at the windshield. A bird strike prompts an automatic flight shutdown while the whole aircraft gets inspected for any damage
that could compromise its safety. They rolled into the hangar,
thinking this inspection would add more than an hour to their
turnaround time.

Todd Taylor, the aviation maintenance technician for the
base's other flight crew, watched from the hangar as the bird was
flung off to the side. Taylor had been trying to think two steps
ahead in order to get the Jayhawk airborne again as fast as possible. Before the helicopter landed he had organized a "hot gas"
refueling, where the helicopter lands, turns around, and remains
on the runway with its engine running while its tanks are filled.

After the bird strike, there would be no hot gas. Now Taylor had another idea. He ran into the storm and searched for the bird. When he found it, a little gray tern with a cap of black feathers, it was 90 percent intact. It had not gotten sucked into the engine or been pureed in the rotor hub. This meant he could order an expedited inspection. Taylor called Clearwater for clearance, which was granted.

Inside, the survivors were unloaded. The rescue swimmer from the second crew had been ordered to meet them and provide first aid. In his zeal, he also thought he'd be replacing Cournia for the second sortie. *Aw, man!* Andrews recalled him saying when told he wouldn't be heading out.

The expedited inspection took only about thirty minutes. The helicopter crew took advantage of the break. They used the bathroom. They drank water and stretched. As ground crew mechanics inspected the helicopter, Andrews played out all the cable and inspected it to make sure there were no frays. He restocked the cabin with chemical lights.

McCarthy, meanwhile, had his own inspection to do: his crew. They had a quick briefing, and McCarthy looked for any signs of exhaustion in his men. He looked in their eyes and listened for slurred words. If he had seen any indication that they weren't ready—if they seemed fatigued, if they'd lost any focus—he could have swapped out his crew for the second unit. But the men were alert. More than that, they were eager to get back out there. And it made sense to keep the same team. They had figured out how to handle both the wind and the waves, and they'd developed a system that was working. As McCarthy put it, they were "in their battle rhythm."

Chapter 11

THROUGH SURF AND STORM AND HOWLING GALE

We're always ready for the call,
We place our trust in Thee.
Through surf and storm and howling gale,
High shall our purpose be.
Semper Paratus is our guide,
Our fame, our glory, too.
To fight to save or fight and die,
Aye! Coast Guard we are for you!

—Coast Guard Marching Song

All Thursday, as Gelera sailed the *Minouche* into the Wind-ward Passage and the Coasties on great Inagua hunkered down in their hooches, Joaquin's power grew and its cloud cover metastasized. Tendril-like bands stained the sky for 185 miles out from the eye. In every direction, the sky was mottled in shades of black, gray, light gray, and even a terrible white that was not the peaceful puff of a cumulus cloud on a summer's day, but the roiling white of ice and water violently sucked up and expelled. The storm itself lumbered south at about five miles an hour, twisting with an amoral anger.

As the crew of the *Northland* sailed west off Haiti's North Claw, they watched the sky fill with increasingly dense gray clouds. The winds picked up and they could feel the long, rolling swells move beneath the hull of the 270-foot cutter. The officers on board had spent the morning and afternoon plotting how to stay out of the storm's way, yet remain close enough to respond to an emergency if needed. By the time they were off Cap du Môle, the wind was blowing 25 knots.

After going over the weather maps, the *Northland*'s captain, Commander Jason Ryan, had decided to sail his ship south, through the Windward Passage, to escape the storm. He wanted to take the ship to Guantánamo Bay, Cuba, using the island as a windbreak. The ship could refuel at the U.S. military base there. He set course.

Even beneath the shelter of Cuba's eastern tip, the seas were eight to ten feet. As Ryan was conducting his nightly routine, checking in on the ship's Combat Information Center, the ship's main command room, then approving the night orders on the bridge, District 7 called. It was about 9:30 P.M. and there was an emergency. They wanted the *Northland* to divert to a cargo ship in distress, the *Minouche*, which had sent a Mayday that the crew was abandoning a sinking ship.

There would be no sleep that night. Ryan gathered his command staff for a quick strategy meeting. The *Minouche*'s location was far enough south of the storm's eye—about 140 miles at that point—that Ryan felt it was safe to sail to it, so they plotted a track line to its coordinates and set a course. They would be sailing downswell, with the wind at their stern. Still, the ship was looking at three to four hours to get to the *Minouche*. That time would be spent scrambling to ready the ship for hurricane conditions.

The captain got on the ship's public address system and announced the mission. "Good evening, *Northland,* this is your captain speaking. We have been diverted by the District 7 Command Center to the motor vessel *Minouche*'s last known position. Prepare for heavy seas."

Ryan's engineer officer started to fill the ballast tanks with water so that the ship would ride steadier in rough seas. The standing order to reduce nighttime engine output was rescinded. As the *Northland*'s twin turbocharged diesel engines roared to life, the first lieutenant ordered heavy-weather lifelines rigged on the forecastle, the forward part of the ship, so that crew members, wearing five-point harnesses, could clip onto the line when they were on deck to avoid being swept overboard. The decks were cleared of unnecessary equipment that might break free in bad weather. Watertight doors and hatches were closed. These were the same procedures as if the ship was getting ready for battle.

Ryan and his officers then debated the rescue options open to them. The seas would be too rough to launch a smaller rescue boat. They might be able to improvise life rings on a long line that could be thrown to the sailors, who could then grab onto the rings and be pulled aboard. The *Northland* could also drape a ladderlike net off the side and try to drift up to the life raft, then have the men jump in the water and climb up. None of these scenarios were ideal.

In his cabin, Captain Ryan dressed in his foul-weather gear. He grabbed his binoculars and an extra flashlight. Then he went to the bridge and tracked their progress. The *Northland* was moving at a fast clip of about 18 knots, aided by the running winds. But as soon as they cleared Cuba and entered the open expanse of the Windward Passage, the *Northland* began to rock.

Joaquin's winds were coming from the south, but also from the west here, and this created a confused sea state. Waves coming from one direction met waves coming from another. Sometimes their energy would cancel each other out. Other times they would combine and stack up, turning fifteen-foot waves into twenty-footers or even higher.

Meanwhile, the *Northland*'s crew were constantly scanning the water ahead with infrared cameras. Floating cargo has a slight temperature differential, enabling the crew to spot debris before the ship crashed into it.

The ship approached the scene at about 1:45 A.M. Ryan made radio contact with the two Good Samaritan vessels, the *Falcon Arrow* and the *Cronus Leader*, to let them know they were free to go. After keeping vigil over the tiny raft, Captain Gokhale could finally continue his journey. (The ships moved off, but stayed in the area, avoiding the storm.) Then Ryan radioed the Jayhawk crew for a status update. At that point, the helicopter had just finished hoisting the fourth survivor. McCarthy advised Ryan that the pilots were wearing night vision goggles, which meant the *Northland* had to be careful not to blind them with its searchlight.

Ryan asked how the *Northland* could assist. Run an orbit around the raft from about a mile away, McCarthy answered. That way the helicopter could use the ship's navigation lights as a reference for the horizon. Roger that, Ryan responded.

As the *Northland* set a course circling the raft, the Jayhawk lifted another two sailors. Then a band of intense rain and wind came through. Ryan and his sailors watched as the helicopter struggled to hold its position, flying up to get above the bad weather. Twenty minutes later the Jayhawk descended and lifted two more sailors, for a total of eight, and McCarthy announced their intentions to fly back to Great Inagua. McCarthy asked the *Northland* to maintain radio contact with the survivors in the raft.

———

From inside the raft's enclosed tent, the men could hear the thud-thud of the helicopter fading as it headed back to Great Inagua with their crewmates. They were now down to four. It was completely dark other than the distant lights of the *Northland*. The assault from the waves continued, but the raft was holding up. Despite assurances from McCarthy that they were coming back to rescue them, despite the ship hovering on the horizon, the men felt utterly alone as the storm—a storm that seemed to be attacking them personally—raged outside. *They are not coming back,* the chief engineer told his captain. *I think we are going to die.*

The bird strike on Great Inagua that delayed the return of the Jayhawk undoubtedly exacerbated the sailors' despair and sense of abandonment, but if the engineer could have seen inside the *Northland* at that moment, he would have been reassured that no one had forgotten about them. Inside the ship's Combat Information Center, an infrared image of the raft was blown up on a forty-six-inch screen. The Coasties were watching the raft intensely.

Finally, about an hour and a half after it had left the scene, the helicopter hove back into position over the life raft with a reassuring whoosh of the rotors. The rain seemed to have intensified, coming down in solid sheets that knocked the aircraft around. Post steadied the Jayhawk into a hover, and Cournia took his stance by the door, ready to deploy. But something was wrong. Post, who couldn't see anything out the windshield, was flying almost entirely by his instrument panel. His altitude was good, but his hover bar showed him moving even though he felt as if the aircraft was stationary. His mind couldn't reconcile how his body felt with what the instruments said. This was a dangerous sensation for a pilot. Vertigo could set in and the pilot could think up was down, and try to fly accordingly. He needed to reestablish some frame of reference. Post alerted his crewmates that he was having trouble staying oriented. He hit the "auto depart" button, which takes control of the Jayhawk and lifts it three hundred feet in the air, then reestablishes an even longitudinal hover. It's a reset, a way for the pilot to start over and try again. From there, Post knew which way was up.

McCarthy, meanwhile, radioed the *Northland* and asked again if the ship would be able to launch a rescue attempt on its own. He needed to constantly assess what options were open to him. But Ryan was even more convinced than before that a ship-based rescue was too risky in these conditions.

Post flew the Jayhawk back down. The raft had drifted, so the crew had to spend a few minutes searching for it. Once they'd located it and positioned themselves above it, Cournia again sat in the open doorway and clipped the hoist hook into his harness. Then he pushed himself out of the helicopter into the night.

In the water, he quickly fell back into the rhythm he had established earlier. He swam up to the raft, pulled himself in, and grabbed the first person he saw. This time it was Captain Gelera—his translator and the one survivor who had specifically asked to go last. "It was dark. I grabbed the wrong one," Cournia would say in retrospect. But at the time, no one wanted to argue and slow things down, so Gelera splashed into the water. Cournia gave the signal for the basket. Andrews lowered it and Cournia loaded Gelera in. Thumbs up. Andrews hit the hoist button and started lifting the basket. Cournia hung from the basket as it lifted out of the water, as he had done on the previous hoists, in order to keep the cable straight and minimize swing. That technique had been working well. But this time, as he let go and slipped back into the water, a gust of wind pushed the helicopter downward, lowering the basket with Gelera in it. At that exact moment, a large wave came rushing forward. The wave swallowed the basket and pulled the cable taut, like a big fish on a line.

Inside the cabin, the soothingly calm robotic voice of the alarm sensor warned "Altitude, altitude," as the Jayhawk dropped and Post struggled to gain some lift. Down in the water, the basket was getting carried away by the wave so swiftly that Andrews worried it would get ripped from the hoist. He began frantically playing out cable to put some slack in the line. Then, afraid he might run out of cable, he started giving the pilot directions to keep up with the basket: "Back and left ten. . . . Back and left twenty." Pilots, as a rule, don't like to hear big numbers. It means something is wrong and getting worse. Post was no exception. He was also dealing with the added challenge of having his sensory perception challenged; the black void outside still played havoc with his orientation, and while he maneuvered the controls to fly backward, he didn't feel like the craft was in fact moving in reverse. The basket, meanwhile, came ripping through the wave, swinging back wildly in the other direction. Gelera clung with both hands to the sides of the basket as he was plunged into a sensory spin cycle, first submerged in seawater, then sent reeling uncontrollably through the dark night air.

From the life raft, Chief Mate Latigo peeked through the curtain and saw his captain swinging in such a steep arc, he was sure the basket would flip and Gelera would tumble into the sea. Inside the helicopter, the violence of the return swing pinned Andrews's head against the door, the cable gouging a deep line in his helmet and knocking his radio off. He was freed only when the cable swung back again in the other direction. By then, Post had regained control of the aircraft, which helped stabilize the cable, allowing Andrews to reel in the basket with a white-knuckled Gelera in it, for the most part unscathed.

Gelera crawled out of the basket and Andrews unclipped it from the cable, then sent the hook down to Cournia for the hover-taxi. But as the cable ran through Andrews's gloved fingers he felt a snag. This was not good. Andrews inspected the cable as best he could in the dim light. At about the seventy-foot mark some strands in the cable had broken, likely as a result of the whipsawing motion during the last recovery. As with the bird strike, an equipment problem of this magnitude triggered an automatic mission shutdown, per Coast Guard regulations. The helicopter would again have to return to Great Inagua, where the crew would have two options. They could swap out the cable drum or they could take the other helicopter. Andrews hoisted Cournia into the aircraft and explained the situation to the crew. Cournia suggested a quick splice, but in Andrews's estimation, the spot where the cable was frayed was too high—too far from the basket connection—and the splice would not be strong enough for the strain of the work they still had to perform: hoisting the last three sailors. There was no avoiding it. They were headed back.

Cournia knelt down next to Captain Gelera and explained what was happening. Gelera, struggling to contain his emotions, desperate not to leave anyone behind, replied: "By the way, sir, thank you so much for helping us. But please go back and save the crew members."

McCarthy again checked with the *Northland* to see if there was any change in their assessment of the conditions to engage in a rescue operation. Captain Ryan discussed the options, which still excluded a small boat rescue. Ryan said they could try to drift close enough to the raft so the men could climb aboard. But to do that he would probably wait until there was some daylight, which was still hours away. A nighttime rescue was just too risky. The two men agreed it was best to continue with the helicopter operation for now, even with this new delay.

As Post pointed the Jayhawk northwest toward the island, McCarthy radioed that they would be returning with a frayed hoist cable. He told the base they needed to use the other helicopter and asked that it be prepped and ready for flight. But in keeping with the night's progress, there was yet another problem. The storm, which by now was heavily battering Great Inagua, had knocked out the island's main power supply, which came from a giant generator. The air base had a backup generator that immediately kicked in, but the storm killed that as well. This meant the massive concrete doors wouldn't open. There was nothing for the Jayhawk to do except continue to the base and hope that the problem would be solved before they arrived.

On base, the ground crew scrambled for a solution. A giant chain connected to four different motors pulled the doors open and shut. Civilians were in the water, and that helicopter had to come out. The men on base had only minutes before the Jayhawk would return. Todd Taylor, the aviation technician who'd expedited the inspection after the bird strike, told his crew to grab tools and dismantle the motors. Once they were disengaged, the doors would roll with less resistance. The men then strapped to the door one end of a sling normally used to airlift equipment. They attached the other end to the mule, the vehicle used to pull the helicopters out onto the runway. Then they gunned the mule's diesel engine, with its 212-feet-per-pound of torque, and wrenched the doors open, inch by inch.

By the time the Jayhawk landed, the ground crew was hooking up the second helicopter to tow it onto the runway. Post, McCarthy, Andrews, and Cournia jumped out, grabbed the gear they needed, and boarded the new Jayhawk. Gelera headed into the hangar to join Maco and the others, finally safe.

The *Northland,* meanwhile, had moved to within about a half mile of the life raft and was sailing an oval-shaped pattern around the last three survivors. The seas were still too rough to get any closer. Joaquin remained a Category 4 hurricane, gusting up to 130 miles per hour, and it had grown in size. Hurricane-force winds now extended fifty miles out from the eye, and tropical storm–force winds extended out two hundred miles, precisely where the rescue operation was taking place. There was a glimmer of hope, though. The storm was beginning its long-awaited shift to the northwest, a very slow turn at about three miles an hour.

These were still dangerous seas for the tiny raft. But there were dangers also for the 270-foot *Northland.* Turning a ship in high seas can be perilous when the ship is broadside to the power and unpredictability of storm waves, and the cutter was doing it regularly—during every rotation of its oval pattern. It required a dexterous hand at the wheel. The conning officer, who gave commands to the helmsman, was trying to time the waves as he gave his commands. The idea was to drive into a swell slowly, then speed up as the ship turned, to minimize roll. At every turn, the quartermaster of the watch would wait for a signal, then announce over the public address, "Stand by for heavy rolls as the ship comes about." The conning officer would then give the command for full rudder to turn the ship and "goose the engines" to drive the ship around quickly. Across the ship, the crew were doing crabwalks to get around, scurrying from one handhold to the next, which is why the ship's safety maxim is "One hand for you, one for the ship."

Eventually, the helicopter crew whirred back onto the scene in the new Jayhawk. Post, confident and practiced in his ability to hover in the high winds now, smoothly maneuvered into position over the raft. Cournia clipped in, and Andrews lowered him into the waves. The crew was synced like gears in a watch, moving quickly with as little wasted effort as possible. And to Cournia, it looked as if the lightning and rain had gotten even worse since the last sortie, which could generate more friction in the blades.

Cournia slipped into the water and swam over to the raft. He pointed at Jules Cadet and pulled him into the sea.

Throughout the night, Cadet had oscillated between bouts of fear and surges of hope. When they'd first jumped into the raft and abandoned ship, he'd thought there was a strong possibility he would die. When the helicopter had arrived and started rescuing people, he'd thought they had a chance. When the helicopter left without him, his mind had spiraled back to its fixation on impending death. Then, when they survived inside the raft for so many hours and the helicopter returned for a third time, he again believed they might survive. But now that he was in the water, with the rescue swimmer gripping him securely, he somehow felt simultaneously safe and accepting that he might *not* live, despite the fact that he was closer to being rescued than he had been all night. It didn't make sense, but he felt a strange mix of happiness and fear. He let the swimmer guide him through the water, which was reassuringly warm as well as powerfully threatening, and let the waves he had been cowering from all night wash over him.

Cournia swam Cadet over to the basket and loaded him inside. As the basket was hoisted, Cournia hung on to minimize swing, and as he was lifted out of the water he felt a sudden jolt pass through his entire body. It was so strong that it locked his arms in a spasm of convulsed muscle. He tried to free his grip but couldn't. Eventually the jolt passed and Cournia was able to drop back into the water.

A special static discharge cable—which had been attached to the basket to siphon excess electricity generated by the helicopter's blades cutting through the air—had somehow ripped off, and a current of electricity had passed from the helicopter's metal frame down the cable to the basket. This meant Cournia had to be extra careful to make sure the basket was in contact with the water whenever he touched it, in order to provide a ground for the electricity.

The second-to-last survivor pulled out of the raft was Henry Latigo. When Cournia pointed at him, Latigo scooted forward clutching a plastic bag. "You can't bring that," Cournia told him. "Please, sir, may I?" Latigo pleaded. "It has our passports and certificates. If we don't have this maybe we can't work?"

Latigo was a lifelong sailor, which in all practical terms meant he was a man of no nation. Like Gelera he had shipped out early, leaving his home on Cebu Island as a young man to become a merchant mariner nearly forty years ago. He had been married, but the union didn't survive his long absences. He hadn't been home to see what family he had left in seven years. He had no permanent address other than the ship he was on, and right now that home was at the bottom of the sea. Without papers, Latigo would be as adrift in the modern world as he was right now. He was a good choice to guard that bag and make sure it arrived onto dry land. It contained his entire existence.

Cournia relented and pulled Latigo and his big plastic bag filled with the crew's documents, as well as the boat's money, into the water. Holding on to the collar of Latigo's life vest, Cournia swam him to the basket, floated him inside, and sent him up to the helicopter. When the cable came back down, Cournia clipped himself in and was taxied over to the raft for the very last survivor.

The last sailor's rescue went so smoothly it was almost forgettable—Cournia, utterly spent by then, has almost no recollection of it—except for the unforgettable fact that a man's life was being saved. At any rate, the sailor rode up without incident—the last in the ragged line of sodden, desperate men who had been on the receiving end of Joaquin's fury all night long. When the sailor was safely inside the helicopter's cabin, Andrews unclipped the basket and sent the cable and hook down one last time. Cournia, bobbing in seas that had tried and failed to subdue him for the past ten hours, grabbed the hook and clipped in. As he was lifted out of the water and hoisted up by his crewmate, he felt his body go limp with fatigue at an effort that had pushed his training to the limit. Around him the winds still whipped and below him the waves still collided and plumed into a white froth. Dawn was just starting to speckle the sky with its light, faint rays of green and pink struggling to make their way through the cloud cover.

McCarthy radioed the *Northland* that the mission was over. District 7 was notified and radioed in one last request: Could the swimmer please puncture the life raft so it would not be a hazard to shipping?

Cournia, safe in the helicopter and finally headed back to base, rolled his eyes.

The request was politely declined.

The storm was still raging, the sky still mostly dark, but McCarthy and Post could see the day's first light bleeding over the edge of the horizon below the cloud cover as the Jayhawk settled onto the tarmac back on Great Inagua. It was 6:25 A.M. In the left ankle pocket of McCarthy's flight suit, where it had been the entire time, was his daughter's stuffed turtle.

After landing and signing the aircraft in, Post walked over to greet the *Minouche* crew, who were huddled under blankets on the hangar deck. He wanted to make sure they were okay, of course, but he was also curious about what had happened on the ship. The captain explained how their ordeal had gone down: the list, the loss of engine power, the big wave. It had been a horrific night, but they were all smiles now. They said thanks—*mèsi* in Haitian Creole—shaking Post's hand and clapping his shoulder. Post wished them well. Then he headed to his hooch.

The helicopter crew was exhausted but still wired from the adrenaline that had coursed through their veins throughout the night. They all wanted to get a message to their families, but the base's Internet connection and phones were down. Cournia had been messaging with his wife right before he was called to the hangar. His last words to her were something to the effect of *Gotta go, there's a ship down in a storm*. Post's wife, Rachel, also a Coast Guard helicopter pilot, was well aware her husband was on a mission in a hurricane and was waiting with growing apprehension to hear whether he was back. Eventually, the base's communications officer relayed a message through to Clearwater, asking the air station to contact all the families and let them know the guardsmen were safe. Finally, the men could get some rest.

Andrews was starving, but once he got to his hooch, he realized he was too tired to feed himself. Instead, he collapsed into bed.

Medics on the base examined the crew of the *Minouche* and were amazed to find that no one was injured. After the exam, the numerous thank-yous, and the extended goodbyes, the men were turned over to Bahamian authorities, who drove them across a landscape of ripped-up trees, torn roofing, and piles of debris to the police station in Matthew Town. They were dropped off in a large garage, where a police officer took their documents for processing.

It was Friday morning, October 2, 2015. The men were beyond exhausted, not to mention ravenous. At around eight o'clock, Gelera approached the police officer in charge. *Please, sir,* he said. *We need good accommodations: beds, fresh water, food.* The officer replied, *Okay, but you will need to pay for that.* Gelera told him they had money. He also said they wanted to get in touch with the ship's agent back in Miami. The police officer told him the power was out all over the island. Phones weren't working and neither was the Internet.

The police escorted the men to a motel where they rented three rooms. Gelera settled in to transcribe the soggy ship's log over to another notebook. Meanwhile, members of the crew went to find some food. They located a restaurant down the street that was cooking chicken and plantains. *Great!* the captain told them. *Now go and arrange for them to cook food for us every day for as long as we are here.*

Before digging in, Gelera pulled his crew into a circle and they prayed—first in Creole, then in Spanish, and then in Tagalog. Afterward, at long last, they had their first meal since before their ship sank.

Meanwhile, the Coast Guard's mission was far from over. District 7's Command Center in Miami and Air Station Clearwater were still trying to figure out what happened to *El Faro*.

The previous day, after they had been alerted about the ship,

Captain Lorenzen and Commander Phy had strategized how they could get eyes on the area. As the day wore on and the storm only strengthened, they had been reluctant to send an aircraft out in the fading light. Before the day was over, they'd committed to sending a C-130 into the storm at first light on Friday morning. They called the pilot who would be on duty then, Jeff Hustace, for a strategy session. This would be a group decision. Sending a C-130 into the eye of a hurricane was not something they would do lightly. Lorenzen and Phy wanted to make sure Hustace felt confident he could do the job.

Hustace, thirty-four, joined the Coast Guard in 2005, after graduating from the University of Nevada and a brief stint working construction. He had been a pilot for eight years. He had never flown into a hurricane before, but had trained over and over to fly in the toughest conditions imaginable. He studied photos of the storm taken by satellites and hurricane hunters and began to plot a path into Joaquin that involved entering from the south side and weaving in and out of the bands. This was doable, he told Lorenzen and Phy.

Then he went to his bunk and tried to get some sleep.

In the aftermath of the *Minouche* rescue, Captain Ryan directed the *Northland* to continue south through the Windward Passage, where they could finally get some shelter from the storm. Then, exhausted from the night's mission, he lay down on the couch in the wardroom and slept for a fitful hour or two. When he woke, he checked his emails and worked on a situation report about the night's activities. In the afternoon, as the ship sailed under the lee protection of Cuba, they had only one large westerly swell roll down on them, rocking the ship. It felt like an appropriate end note to a hectic and stressful night. But Joaquin wasn't done with the *Northland* yet. The storm was still churning, and there was unfinished business out there.

Around five thirty on Friday evening, the Command Center

at District 7 contacted the *Northland*'s Combat Information Center with a request that the ship head north to *El Faro*'s last known position. The CIC watch supervisor forwarded the request to Ryan. Joaquin was now about fifteen miles northwest of San Salvador Island, downgraded slightly to a Category 3, but still blowing 125 miles per hour in a 50-mile radius, with tropical storm–force winds extending out 205 miles. Ryan did not like the idea of sailing through the dark in high seas and heavy winds to a location where, if *El Faro* had sunk or even simply lost cargo, large metal shipping containers could be floating partially submerged. He told the watch officer to decline. Then he went down to dinner. The watch officer returned a few minutes later. District 7 wasn't giving up. They wanted to talk to Ryan on the satellite phone.

Ryan made the call, and once again declined to sail into the hurricane. Captain Coggeshall and his team in District 7 tried a compromise. Could the *Northland* sail just south of Crooked Island and wait there until the storm moved away? Ryan studied the charts and concluded that this would be pointless. The longer the wind blows, the bigger the seas get, and Joaquin had been parked in the area for more than two days now. He explained that the *Northland* would spend the night battling the waves, and by morning both ship and crew would be rag dolls running out of fuel. However, Ryan said, the storm was already beginning to weaken and slowly move north. By morning, he'd feel confident taking his ship up to the islands.

Finally, they all agreed that the *Northland* would head to the U.S. base at Guantánamo Bay to refuel, then sail to the Bahamas in the morning. But when Ryan called the base's commanding officer to request permission to refuel, he was declined. Forty-knot winds (46 miles per hour) were blowing pier-side, and the commander didn't think they could safely dock the ship. Ryan had no choice. He would have to keep the *Northland* at sea overnight.

The next morning, the *Northland* pulled up to the base's

dock at about 7:30 to refuel. Crew members jumped off to quickly email and call home using the base library's Wi-Fi. That's when the commander of the Coast Guard's Port Security Unit, James Hotchkiss, asked Ryan if he had any updates on *El Faro*'s status. One of the security unit's members, a Coast Guard reservist named Richard Griffin, had a brother, Keith Griffin, who was a crew member on *El Faro*, Hotchkiss explained. Richard had arrived at Gitmo a few days earlier to do his reserve duty and hadn't even unpacked when he'd heard that his brother's ship had lost communication. Richard was already on his way back to Jacksonville, Hotchkiss said. He asked Ryan if he would please keep him informed. Of course, Ryan said.

It took several hours to fill the ship's fuel tanks, but by 1:00 P.M. the *Northland* was ready to depart. As the ship sailed out of Guantánamo Bay en route to the Windward Passage, it passed the Coast Guard cutter *Resolute* coming in to refuel. It would be right behind the *Northland* as the ships sailed north to try to solve the mystery of what had happened to *El Faro*.

Chapter 12
CLIMBING THE CLOUD DECK

By late Friday, TOTE employees and officials with the Seafarers International Union and the American Maritime Officers union had begun the terrible protocols of notifying next of kin that their loved ones' ship was missing near a hurricane. This entailed going through the employee list of emergency contacts, which, depending on the sailor and how diligent he or she was about maintaining this kind of thing, might be years old. Some of the contacts had moved. Others were simply traveling or out of town. And when they were reached? What then? They had to be told, in calm tones, that the company had lost contact with *El Faro* and that the Coast Guard was preparing to search for the vessel.

Chief mate Steven Shultz's wife, Claudia, was in Brazil with her sick father when word finally wound its way down to her. Able seaman Jack Jackson's brother's house had been damaged during Hurricane Katrina, so Jack had put down his brother's girlfriend's landline, which was not working at the time. Jackson's sister had recently moved to Florida to be closer to him, so her contact information was old as well. Eventually, though, word got where it needed to go. People heard things, or saw the news, and called in. Then they started heading to Jacksonville to seek some answers. And to wait.

Tina Riehm, wife of third mate Jeremie Riehm, was working at the daycare center she owned when a woman from the AMO called to tell her the company had lost contact with Jeremie's ship and were working to reestablish communication.

Tina thanked the woman and didn't put too much thought into it. Ships at sea lose communication.

But when she got home, turned on the news, and saw a story about a ship lost in a storm, she called the AMO back. *Tell me everything*, she said, *and keep me informed*. She had a terrible night and woke up sick as a dog. In her heart, she knew Jeremie was trying to tell her something important.

Lieutenant Commander Jeff Hustace went to bed on Thursday night in Air Station Clearwater's crew quarters not knowing if the mission to search for *El Faro* was a go. His answer came at 6:00 A.M. on Friday, October 2, over the loudspeaker: "Put the ready C-130 on the line. First-light search for steam vessel *El Faro* in the vicinity of Crooked Island." That was his plane, and it was time to fly into the storm.

Hustace and his crew, including copilot Brian Farmer, reported to the flight-planning room inside the C-130 hangar, where they stood at a belly-high table covered with maps. The men at the table were plotting the ship's last known coordinates.

"How do you want to approach?" Phy asked Hustace.

Hustace reviewed the updated satellite images of the storm and stuck to his plan to enter from the south side. That approach would provide the best entrance points because the winds were weakest there—"weakest" being a relative concept for a Category 3 hurricane. Phy listened to his pilot, and what he heard made sense.

The men talked out some final logistics, folded the maps, and headed to the Herc.

Sending a plane into a hurricane on a search and rescue mission (as opposed to flying above or through the storm for research, as a hurricane hunter does) is not common practice. In all his decades serving in the Coast Guard, Captain Lorenzen doesn't recall ever doing it before. It made him extraordinarily nervous. It was, perhaps, the toughest decision he had ever had to make in the line of duty. But he trusted his men and the equipment. If the pilot and crew believed they could do it, he would not second-guess them. And the job needed to be done. So when Phy called him for the okay to launch, Lorenzen gave the go-ahead.

Hustace was comfortable with the mission; he knew the Herc, its range, its power, and its capabilities. He knew the plane carried lots of fuel, and he had flown in heavy weather before. But his commanders were anxious, and he kept having to reassure them that he would be careful.

The crew of eight climbed into the aircraft and strapped in. Hustace dropped the condition lever to "run," pulled the starter switch, and the four massive turboprop engines began to spin. Then it sped down the runway and lifted off over Old Tampa Bay. The plane flew south across the state toward Fort Lauderdale at about twenty thousand feet. Just off the shore, Joaquin's cloud deck became visible. The south face, which had looked reasonably approachable on paper, was now occluded by a mass of dark clouds. The navigator noticed embedded thunderstorms on his radar, and the two pilots, who were wearing night vision goggles because it wasn't completely light yet, saw lightning crackling through the clouds. Hustace called the Air Force hur-

ricane hunter, which had just flown above the storm. Its crew
confirmed the embedded lightning and storms on the south
edge. They further advised that the bands on the north side were
a lot cleaner. The winds were stronger there, yes, but overall it
was safer.

So Hustace scrapped his plan, changing course and flying toward the north face. As they rounded Joaquin, flying under the
cirrus streamers on the edge of the storm, the crew could see the
distinctly stratified layers of cumulus and cumulonimbus clouds
along the vertical cloud deck. Hustace picked an altitude between two layers in the cloud deck's formation, at about nine
thousand feet, and flew into the feeder bands, dark clouds with
what looked like hanging curtains of precipitation under them.
The plane shot through the bands into gaps where they could see
the ocean beneath them. The crew marveled at how well organized the storm was.

The idea was to get to the "search box," which included *El
Faro*'s last known position and the area around it into which the
ship might have sailed or drifted. Once inside the box, the crew
would do a radar search.

The plane shook and trembled in the winds. Whenever the
aircraft passed through a feeder band of clouds and rain, the
lashing water temporarily blinded them. As the wind swirled
around them, sudden gusts would shove the plane, traveling at
about 160 to 170 knots (184 to 195 miles per hour), forward 50
knots at a time. Downdrafts would drop the plane without warning. Then they'd break through the feeder band to the calm
outside of it.

Eventually, after about forty minutes, they made it to the
search box. But at nine thousand feet, the rain and clouds were
too thick for the radar to "paint" the surface, the term the men
used for the picture created by sending electromagnetic waves to
ping off solid objects and bounce back. They had to fly lower and
try again. Hustace piloted the plane out of the storm, descended

to about two thousand feet, and flew back in, again winding through the gaps in the storm's feeder bands. (The men didn't know it at the time, but the winds were shaking the plane so hard that some of its fastener bolts were coming loose.) This time it worked.

The crew deployed the C-130's full arsenal of detection: surface radar, Selex radar, weather radar, and the CASPER (for C-130 Airborne Sensor with Palletized Electronic Reconnaissance), which used color graphics to reveal the surface in a variety of hues. On this mission, however, the CASPER was set to black and white because it strengthened the color contrasts. The plane's nose had two radar turrets. The crew scanned back and forth, pulling images up on their screens for closer inspection. The images were clean enough that they were able to detect big rocks on the shore of the islands. At the same time, the pilot and copilot scanned the surface of the ocean with their eyes.

The C-130 continued this process as it methodically covered the search box, flying out of the storm when necessary, then using a sort of zigzag flight pattern to navigate back in. At one point, Hustace took them to within about fifteen miles of Joaquin's eye, which was as close as he dared. Here, sudden downdrafts dropped the plane eight hundred feet at a time. Hustace would then "feed in a bunch of power" and climb back up again. These were the hardest sustained conditions he had ever flown in, yet his faith in the Hercules never wavered. But there were limits he had to be aware of. One of them was fuel. Due to the weather, they were burning about 6,500 pounds of fuel an hour, about 1,500 pounds per hour more than usual. The other limiting factor was the Coast Guard's strict crew rest requirements. The crew was only allowed to fly eight hours before it had to be on the ground.

After being in the air for about seven and a half hours, the crew had no choice. Hustace flew the plane out of the hurricane, which was like stepping through an alternate dimension. One moment they were surrounded by darkness and chaos; the next, peace and light. Rather than flying all the way back to Clearwater they headed for the closest Coast Guard landing strip, which was in Opa-locka, Florida, near Miami. After the plane touched down, the men stiffly exited their aircraft with mixed emotions. They knew they had done their job as thoroughly as possible. If a nearly eight-hundred-foot ship had been afloat down there, or anywhere near the surface, they would have seen it. Yet they were frustrated. They had found nothing that could ease the minds of the worried family members gathering in Jacksonville.

As the ground crew in Opa-locka inspected the plane, they discovered the loose bolts that had been shaken free during the intense turbulence, plus a small fuel leak caused by the winds. As a result, the plane was grounded, taken out of commission.

This was a significant blow to the search effort. In District 7's command center, Captain Coggeshall, the search and rescue mission coordinator, was frustrated to learn about the operational loss of the C-130 in Opa-locka. He needed that plane in the air! Coggeshall reviewed the assets he had at his disposal. Coast Guard Air Station Miami had an HC-144—a medium-range, twin-engine airplane—and a few MH-65 Dolphin helos, the Jayhawk's smaller cousin. Air Station Clearwater had its regular roster of four C-130 Hercules, plus two C-130Js that had been moved south from Elizabeth City, North Carolina, to keep them out of the forecast track of the storm. There were also the Jayhawks in Clearwater, Great Inagua, and Andros. Finally there was an additional Jayhawk in Savannah, Georgia, just above Jacksonville. It was already midafternoon. If the Coast Guard launched another C-130 from Clearwater, it would get to the search area just as the sun was setting, which made no sense. And neither the Dolphins nor the HC-144 had the range for a sustained search in heavy winds.

Given the dwindling daylight, Coggeshall decided to ask a Jayhawk from Great Inagua to fly up along Crooked and Long

Islands, looking for anything—signs of life, survivors, life rafts—and try to get as close to *El Faro*'s last known position as possible, even though the eye of the hurricane was still in the same general area.

Lieutenant Rick Post woke up in his hooch on Great Inagua on Friday afternoon. His body ached and he was still numb from exhaustion. He stumbled out of his cot and made his way to the kitchen for a cup of coffee. The pilot for the other Jayhawk crew, Kevin Murphy, was suiting up, preparing to start his shift. Post vaguely recalls the words "hurricane" and "*El Faro*" being spoken. His body recoiled. He grabbed his coffee and slumped down at the table, grateful he was not the one heading out this time.

Murphy, on other hand, was eager to get up in the air. He had been expecting this since Thursday, when they'd first heard that *El Faro* was having engine trouble. Murphy was proud of his colleagues McCarthy, Post, Andrews, and Cournia. They had saved lives. Now, he thought, it was his turn. Murphy, a lieutenant, had been an aircraft commander for only about six months. His co-pilot was even more junior. This would be their first time taking part in a search mission of this magnitude, let alone in a hurricane. During the mission brief, Murphy and his crew were given the coordinates for an EPIRB signal from a container on *El Faro*. (Unlike the one on the ship itself, the EPIRB on this container was encoded with GPS.) It was solid evidence of the ship's presence and an important clue.

The conditions on Great Inagua had improved slightly overnight. The rain was still intense, but the wind was blowing at only 40 knots, and there was plenty of light to see by. The crew strapped in and lifted off, and Murphy pointed the nose northwest toward the center of Hurricane Joaquin and Samana Cay, the last known position of *El Faro*.

As they flew, the cloud cover increased and the visibility decreased. The rain intensified. The wind was now blowing about 65 knots, and the cloud ceiling sank to about 1,200 feet. Even at that height, they had trouble seeing the water. Murphy dropped the Jayhawk down to about 700 feet. Just past Acklins Island they reached the storm; 103-mile-per-hour winds hit their port side, forcing the helicopter to make a kind of crablike progress forward by pointing the nose into the wind at an angle. The rain came down in a sudden torrent, plastering the windshield in sheets of water. Eventually they made it to the latitude and longitude given by the EPIRB on the container, but the crew didn't see anything on the water's surface. Murphy flew around the area, a tricky maneuver in the wind. Flying at 100 knots, the helicopter would turn into the wind, which would slow them down to only 10 knots. Then, as they came back around, with the wind behind them, the helicopter would zoom forward at 200 knots.

Murphy and his crew had faith in the Jayhawk; it was an incredibly tough and stable bird. But the discrepancy between what they saw outside and what the instrument panel read messed with their minds. They didn't feel the movement, but when they looked outside, it appeared as if the Jayhawk were being blown sideways at 100 miles per hour. The difference between the visual cues and the sensory input was the most disorienting thing about flying in these conditions. Below them, chaotic thirty-foot waves rose and fell in an angry dance. Still, they were only a few hundred feet above the water, and Murphy was confident they would have seen a life raft or a survivor below them.

After a few hours, sunset approached. The sky was getting darker, compounding the storm's disorienting effects on the Jayhawk crew. Murphy reported to SAR mission control the results of their search and then headed back. It took them two and a half hours to reach their search area, but with the wind at their backs, it would take only about ninety minutes to fly back. At their closest, Murphy had piloted the Jayhawk within forty miles of the eye of storm.

The evening of Friday, October 2, descended without any new clues about *El Faro*'s fate or its whereabouts. In Miami, Coggeshall focused on expanding the search zone. He was frustrated that the searches so far hadn't turned anything up, not even a stray shipping container, but he was looking forward to building a search pattern that would effectively scour the area. It had been more than thirty-six hours without any sign of the vessel. In his mind, the likelihood that the ship had sunk grew. This meant the search would now increase its focus on smaller objects in the water. But there was still a chance the ship had lost communications along with power, and was being pushed north by the hurricane.

Coggeshall pulled his team into the conference room that night at shift change. The weather forecast indicated that the storm would move west slightly and slowly crawl north, at about three miles an hour. That gave the Coast Guard an opening to push more assets into the area. The software program the Coast Guard used to plot drift, called SAROPS, for Search and Rescue Optimal Planning System, was not designed for such a big vessel in such extreme conditions. For vessels, the SAROPS maxed out at about three hundred feet. Anything bigger and it was presumed you'd be able to see it with the naked eye or with sensors. Wind speeds on SAROPS topped out at 40 knots.

So Coggeshall and his team resorted to old-fashioned hand calculations. Objects in the wind drift at a roughly forty-five-degree vector downwind. To adjust this for the *El Faro* search, they had to take the maximum 40-knot wind-driven drift (Coggeshall calculated the wind would be a bigger factor than the ocean current in a hurricane) and multiply it by three, then apply it in a northeast direction. Coggeshall then had his team lay out three C-130 search boxes. The first was right over *El Faro*'s last known position. The other boxes tiled out in the direction the hurricane's winds were moving. The C-130s would fly low over the water, searching for lifeboats or individual survivors. Above them, a U.S. Navy–supplied P-8 Poseidon, used for long-range reconnaissance and intelligence, would fly at a higher altitude to look for the ship itself. This would be their plan for the following day. The Command Center would stand watch overnight, but there was little anyone could do in the dark except listen on the radio for signs of life and type messages into the computer to broadcast over the emergency channels.

Coggeshall headed home at 8:00 P.M. to try to get some rest, his mind overrun by the complexity of the search logistics ahead of him.

Meanwhile, the hurricane hunters were still up in the air, thousands of feet overhead, issuing callouts into the empty sea beneath them, hoping to catch a radio signal, a voice, anything indicating the vessel might still be afloat.

Chapter 13

"TOGETHER, AS ONE CREW"

Hurricanes usually blow through a specific location, barreling west or north, spending only hours in one place. Joaquin went south, then sat for days around the Bahamas. As the storm, now a 125-mile-per-hour Category 3 hurricane, churned 165 miles northeast of San Salvador on the morning of Saturday, October 3, 2015, a low-pressure system over the southeast United States drew a steady stream of moisture from Joaquin's upper reaches, drenching the region. There was historic flooding in Charleston and Columbia, South Carolina. Meanwhile, islands throughout the Bahamas were flattened by winds and a deluge. Seventy percent of Crooked Island, Long Island, and Acklins Island were flooded with five feet of water or more. No one, though, lost their life to Joaquin in the Bahamas.

The storm began to move slightly as the day wore on. But this was not the break everyone involved in the search and rescue mission was hoping for. Even as it slowly lurched north of the Bahamas, staggering and stopping like a mean drunk looking for a fight, Joaquin was gaining strength. The low over the southeast United States, and a second low northeast of Joaquin, had dissolved the Atlantic ridge that was inhibiting the storm's northward movement. With the ridge gone, the storm's pace quickened, helping it grow stronger.

By noon a hurricane hunter had measured the storm's maximum sustained winds at almost 155 miles per hour. Joaquin was flirting with Category 5 status. And the area where *El Faro* had broadcast its last message was still inside a zone of dangerous weather.

Nonetheless, the Coast Guard C-130s and the Navy P-8 Poseidon flew the search patterns Coggeshall and his team had laid out the night before. The most northerly box was still too close to the hurricane's eye for the planes to be effective. But the weather had significantly improved for the middle box's aerial search. Visibility there was now about two miles, and the winds were blowing at a comparatively breezy 60 miles an hour. The P-8 was able to scan hundreds of miles at a time on radar. Unfortunately it found no evidence of *El Faro*. The C-130 crews, flying closer to the water, began to see specks of debris floating beneath them. In the afternoon, the crew of the C-130 flying the middle search box spotted three life rings and quickly relayed their position to a Jayhawk helicopter, which swooped over and was able to pick one up. Emblazoned on them were the words "El Faro." It was the strongest sign yet that the ship was not simply disabled and adrift.

When news of the life ring reached Coggeshall back in District 7, his suspicions mounted that the ship had gone down. Of course, it was entirely possible that the life rings had blown off the deck. They would need stronger evidence, such as items that would have been only inside the ship, before they could conclude that *El Faro* was underwater. Coggeshall continued with the plan for the P-8 to search exclusively for *El Faro,* and the C-130s and Jayhawks to focus their efforts on lifeboats, life rafts, and survivors in the water.

The P-8 Poseidon followed a track line that took it to within one hundred nautical miles of the storm's center, offset by one hundred nautical miles to the southeast, and then back. In this way, the plane would be able to use its powerful radar scanners to scour a five-hundred-mile swath of ocean. But by the time it landed later that afternoon, it had not detected anything the size of *El Faro* floating on the sea's surface. It was the C-130s, flying low over the water, that started spotting items in the waves: odd bits of flotsam not immediately identifiable from the plane's altitude of a couple thousand feet. But as the Herc crews continued flying their pattern, it became increasingly clear this was a debris field of some sort, and that it was pluming out for miles in the water below. With this information, Coggeshall and the rescue mission's coordinators felt confident that they could inform the families that it was the belief of the U.S. Coast Guard that *El Faro* had gone down—and that the search effort now would shift to lifesaving equipment and survivors.

That evening, Coggeshall initiated a conference call with the families, a protocol that the Coast Guard would keep in place throughout the search as a way to keep them updated. The service ended up holding daily conference calls at about 10:00 A.M. from District 7's Command Center to inform families of any overnight developments, followed by an early evening in-person briefing at the Seafarers International Union (SIU) headquarters on Belfort Road, in Jacksonville's Southside, a well-manicured neighborhood of office parks.

At the moment, though, there was a frustrating shortage of facts to share with the families—mainly just the Coast Guard's observations from the air. It was more crucial than ever to get ships in there to search close to the water. Twenty-five-to-thirty-foot seas and high winds were still churning the waters on Saturday, making it too rough for vessels. However, this was changing by the hour as the storm continued heading north. In fact, by Sunday morning Joaquin had left the area.

After sailing northwest through the night at a brisk 18 to 20 knots, Captain Ryan had timed the *Northland*'s arrival into the search area to coincide with first light that morning. This was a safety issue; he didn't want the ship sailing through a debris field in the dark. Right on schedule, the ship arrived in the Crooked Island Pass at about 7:00 A.M. Captain Ryan was astonished at how calm the area was, with barely any wind blowing. The sun shone brightly in a cloudless sky. The sea lay flat and lifeless, as if the storm's strength had pulled all energy with it. For the sailors floating into this depleted scene, it was an eerie contrast to what they knew had happened here the day before. But Ryan was aware that as storms move away they suck the clouds with them, and as the wind direction changes, it knocks the seas down quickly.

Soon the sailors smelled fuel in the air. Below them, the water was dotted with chunks of man-made debris, including big pieces of insulated metal that looked like they had once been part of a shipping container. The *Northland* reduced speed, its bow pushing water and scraps of flotsam aside as it continued sailing up to *El Faro*'s last known position. Two tugboats hired by TOTE were already on the scene, beginning their own search pattern, as was *El Faro*'s sister ship *El Yunque*. Ryan made contact on the VHF radio, quickly establishing the *Northland* as the coordinator for the search effort. *El Faro* may have been TOTE's ship, but the Coast Guard would be managing the search.

In short order, sightings of containers and flotsam started pouring in over the radio. It was pretty clear these were the tattered remnants of *El Faro*, given that there had been no marine traffic here the day before—except for *El Faro*. Ryan had beefed up the watch in the ship's Combat Information Center, and the crew there scrambled to log all the information coming in: a group of about twenty floating containers at this coordinate, pieces of twisted steel and a capsized lifeboat at this one, over here a deflated life raft. The *Northland*'s officers pushed the significant sightings to District 7. Recovering all that debris would have been impossible and would slow down the search effort, so only items deemed critical to the investigation, such as lifesaving equipment, were recovered.

The ships were in the middle of a debris field that stretched for forty miles. Ryan and his sailors saw more flotsam torn from shipping containers, as well as pieces of metal that resembled the inside of a bulkhead wall. They spotted two intact shipping containers. The *Northland*'s crew lined the rails with binoculars raised, scouring the water. Beneath them the detritus of life on a ship floated past: plastic laundry baskets, flip-flops, medicine bottles, whole oranges, bags of garbage. More scraps of metal floated by, buoyed by air trapped inside, or insulation attached, including one piece that could have been part of a ship's superstructure. They found a Carolina skiff that had probably been blown off one of the islands. Two bright orange survival suits were found floating in the water as well.

Overhead, the C-130s and P-8 continued their search, aided by Jayhawks from Great Inagua and Andros. The *Northland*'s Combat Information Center coordinated the investigation of debris sightings from the planes, dispatching helicopters and boats to move in for a closer look. Inside the *Northland*'s CIC and bridge, the radios crackled with updates from ships and planes, all of it assiduously logged. Ryan knew the higher-ups at District 7 would have a voracious appetite for information.

In the late afternoon, a Jayhawk was flying a search pattern when the copilot spotted something orange in the water. The pilot descended to three hundred feet. It was a Gumby suit, and it appeared there might be a body in it. The crew's rescue swimmer clipped into the cable hoist and was lowered into the water, where he detached and swam over to the suit. He got within about three feet and saw that there was in fact a body inside, which would have been floating now for almost four days. The swimmer would later describe the remains as "very deformed, unrecognizable." He couldn't even determine if the corpse was male or female. He looked up to the helicopter and drew a finger across his throat. The flight mechanic then sent the cable back down to hoist the rescue swimmer.

Inside the helicopter, the crew radioed the *Northland* that it had discovered a body and would await further instructions. There was a long silence. Onboard the *Northland,* officers were discussing with District 7's Command Center whether to recover the body. While it might seem shocking that the guardsmen would consider not retrieving it, in a SAR situation like this, it's actually a complicated decision. If they did bring the body inside, the helicopter would be considered a biohazard site and would have to return to base and be cleaned. This would take it out of commission for the remainder of the day's search operations.

But before a decision had been made, the Navy P-8 radioed in. Its crew had seen what looked like a person in an orange survival suit in the water, waving at them. The Jayhawk crew received orders to put a marker buoy on the body they'd found, which would transmit the location in the SAROPS program, and fly to the location the P-8 had identified, about fifteen minutes away.

To everyone's deep disappointment, however, it turned out to be a false sighting. When the helicopter arrived, the crew found only some orange fabric from a deflated life raft floating and twisting in the waves, which from above looked like a survivor signaling from the water. To compound the Jayhawk crew's disappointment, when the pilot steered the helicopter back to the general location of the deceased sailor, the data buoy wasn't functioning, and the body had disappeared.

Back at District 7, Coggeshall's team began trying to make

sense of all the incoming evidence. If *El Faro*'s crew had been able to abandon ship in an orderly manner, they would have tried to stay together in the storm. It was also likely that they would have zipped into their survival suits. You can't swim in the suits; you can only drift in them, so it stood to reason that other deceased sailors would have been found near the dead sailor that the Jayhawk found. This set of circumstances, plus the fact that a capsized lifeboat with its sides and bow crushed had been found, along with a life raft, not fully inflated and wrapped in line, led Coggeshall to conclude that the crew had not been able to evacuate the ship in an organized fashion.

All this informed how the next day's search would unfold. While Coggeshall considered this to be on the far end of optimistic, the crew might have made it to the second lifeboat, or another life raft. Still, he couldn't figure out how they would have been able to get from the deck of the ship to the life raft or lifeboat in thirty-to-forty-foot seas, as the waves lifted the life raft and dropped it violently while tethered to the ship. And if other sailors had entered the water in their survival suits, well, they all knew about the deceased sailor.

That evening's debrief was the hardest one yet for the families to endure. Wives, brothers, sons, and daughters were clinging to hope like a psychic shield, something to protect them from the impending knowledge that their loved ones were gone. Coggeshall and his team prepared PowerPoint slides every day to update the families on the Coast Guard's search efforts. The slides listed the number of ships and aircraft used, the square miles they'd searched, and the things they'd found. That Sunday evening, amid the report on the 96,680 nautical square miles scoured that day, they had to explain that a rescue swimmer had found a sailor's remains floating in a survival suit, but had not recovered it.

Family members were outraged. *How could they abandon the body?* In the Jacksonville SIU hall that evening, more than one

hundred people packed the room. The mood had been deteriorating for days. At that moment, they couldn't understand the
Coast Guard's rationale for leaving the body behind. Actually,
they could *understand* the rationale—that the helicopter was
called away to search for a potential survivor, and that even if the
crew had picked up the body, the Jayhawk would have become a
biohazard site that compromised the search effort. It was simply
that many of them could not emotionally process the information.

"As a maritime family, you get the concept of lost at sea," said
Carla Newkirk, the daughter of able seaman Larry Davis. "But
there were *remains*. The consensus of the room was, we couldn't
understand why the person wasn't brought back to us. All of us
believed it was our person."

And that, she added, was "tragedy on top of tragedy."

Monday broke clear and calm, perfect for the search effort. The
Coast Guard was able to flood the skies and water with assets.
The 210-foot cutter *Resolute* and the 154-foot *Charles Sexton*
were now on the scene. The *Resolute* was tasked with patrolling
a smaller debris field about sixty miles away, on the northeast end
of the search area, where it spotted a group of about twenty
partially submerged cargo containers. At one point, a group of
Minnie Mouse dolls that were part of *El Faro*'s cargo floated by
the *Northland*.

The *Northland* coordinated as many as five aircraft in the sky
while also sending out smaller boats to investigate debris. There
was a lot of safety-orange material in the water, and each sighting
sparked a surge of hope that it might be a survivor floating in the
water. The sense of urgency running through the search operation was at a high fever. The *Northland* continued cruising the
debris fields even at night, picking its way through very slowly
with its searchlights on. The calm seas made this less risky now,
and everyone felt that every second had to be used.

Tuesday was another clear, bright day. The sailors on the *Northland* had been working relentlessly for forty-eight hours. On the ships, everyone was either eating, sleeping, or on watch. Captain Ryan could see the fatigue setting in among his crew. They were emotionally drained, staring into the water day after day looking for signs of life, but finding only evidence of destruction. Ryan, pulling on the lessons from his SAR school training, circulated around the ship, reminding the crew that they were a survivor's last advocate, their last hope. The families were relying on them.

By the end of the day, the Coast Guard's command staff had started discussing among themselves the prospect of ending the search. Coggeshall, for one, was confident the search patterns were accurate because the drift calculations had been corroborated by the debris fields. Those debris fields had been thoroughly combed over; the coasts of the islands had been scoured from above. Six days had passed since *El Faro*'s last transmission. They weren't going to find survivors, Coggeshall concluded, only more debris.

Inside District 7 headquarters, the Coast Guard had set up a room for TOTE's representatives to work, making them easily accessible to answer the Coast Guard's questions and allowing them to communicate with their own ships in the search area. At the end of Tuesday, Coggeshall and some of his colleagues approached TOTE with their decision. The TOTE officials, in turn, asked if the Coast Guard would agree to deliver this news in person to the families assembled in Jacksonville.

Tuesday night was a truly sleepless one for Coggeshall. They had all worked so hard, for so many days, and now it was ending in failure. Not only had they failed to find survivors, they hadn't even been able to recover any remains that might bring the families some sense of peace and closure. He had formed a bond with these families—not because he'd gotten to know them especially well, but because he was trying to help them. Now he would have to tell them it was over.

"Our stated goal is to save lives and property," he said stoically. "When you don't succeed, the costs are very high."

In the morning, Coggeshall dressed in his "tropical blues," dark blue pants and light blue short-sleeve shirt with shoulder boards and ribbons on the chest. He put on his white formal uniform hat, which bore an eagle on the front and a leaf cluster on the visor. Then he boarded a twin-engine HC-144 Ocean Sentry at Coast Guard Air Station Miami, along with District 7's commanding officer, Rear Admiral Scott Buschman, and the chief of response, Captain Mark Fedor, bound for Jacksonville.

By 11:00 A.M., roughly two hundred people had filed into the auditorium inside the SIU building and taken seats in metal chairs arranged around a center table, where the Coast Guard officers were waiting. Coggeshall looked out at the crowd sitting before him and was overwhelmed by the magnitude of what he was preparing to tell these folks. He had delivered next-of-kin notifications before, but usually to only one or two people at a time. He and his fellow officers were about to tell a union hall full of men and women that the Coast Guard would suspend its search at the end of daylight today, Wednesday, October 7, 2015. He was saying their loved ones were dead.

In his somber presentation, Coggeshall laid out the timeline of the search: the Coast Guard and Navy ships, helicopters, and airplanes used, what was found, how many miles were searched and where. Then he told the room that in his professional opinion they were unlikely to find any survivors or remains at this point. He was making a recommendation to the district commander to suspend search efforts at sunset. Next the rear admiral spoke, announcing that he was accepting Coggeshall's recommendation and had given the order to suspend the search.

Despite the passing of a week, despite the overwhelming physical evidence that the ship had sunk—the lifeboats, the life rings, the Gumby suits—many of the families had not allowed themselves to accept the reality of what had happened. As long as there was no body to weep over, hope remained. Now the Coast Guard was telling them all hope was gone.

Carla Newkirk sat there in stunned disbelief. Until that moment, she'd remained certain in her heart that her father was alive. When the other family members around her had succumbed to despair, she'd blocked them out, not wanting to hear it. Her father was as solid a presence in her world as the ground she walked on. Throughout her life, her daddy had gone to sea, first as a fisherman, then as a merchant mariner. He had always returned. Not returning was as inconceivable to her as the tides failing to change. Larry Davis went to sea and came back with stories of big fish and bigger waves, and his daughter never believed it was dangerous. It was simply his job. The Coast Guard was giving up far before she could.

Jack Jackson's brother, Glen, listened in on the phone from Louisiana, stunned like Newkirk but also angry. *How was it possible,* he wondered, *that in 2015, with all the satellite forecasting and communication technology available, that thirty-three people on an American ship could die at sea? How was his brother allowed to work on a boat with an open lifeboat and an EPIRB that didn't have GPS?* His brother had crossed every body of saltwater on the planet, only to die on a Caribbean milk run just a few hundred miles from home.

Tina Riehm did not attend the event in Jacksonville, either. The timing was unintentionally cruel; it was her daughter Clancy's twenty-second birthday. Tina had to tell friends and family to ignore Clancy's birthday: no presents, no cakes, no cards. All week they had been frozen at their home in Pine Island, Florida, unable to leave, caught somewhere between grief and hope.

Richard Griffin's reaction was: My brother? *Gone?* The guy who could fix anything and figure out how to solve any problem, gone? That didn't make sense. Richard couldn't get the image out of his head of his brother, Keith, holed up in some Bahamian bar, drinking beers. And in the swirl of his confusion, he grew illogically angry. He knew better; he was in the Coast Guard, after all. But he also knew his brother, and how resourceful he was. Richard was not ready to give up on him yet.

But the Coast Guard, while operating with empathy, was also operating with facts. They had expended all the resources a modern nation could muster and had come up with nothing. Their decision was final.

After leaving the SIU hall, the officers drove over to TOTE's offices and made the same announcement to another two hundred people.

Then the HC-144 flew back to Miami, where Coggeshall oversaw the closing of the case. The last C-130s returned to base from that day's search. The *Northland, Resolute,* and *Charles Sexton* ended their search patterns and were released to set course for their next assignment.

Inside the Command Center, the staff closed the files in the MISLE, or Marine Information for Safety and Law Enforcement, database. That night, the operations specialist did not type a message into the emergency computer system, and no planes transmitted callouts to the phantom ship.

News that the search was over resonated across the country. Throughout the week, the disappearance of *El Faro* had been on the front page of *The New York Times, The Washington Post, The Wall Street Journal,* and all over the cable and network news. Newspapers and TV stations from Maine to Florida had profiled the local sailors and the anguish of their families. What had been a maritime mystery was now officially a national tragedy.

At 6:18 P.M., President Barack Obama released a solemn message from the White House:

> "The captain and crew of the *El Faro* were Americans and Poles, men and women, experienced mariners and young seamen. They were beloved sons and

daughters and loving husbands and fathers. They were dedicated engineers, technicians and a cook. And these 33 sailors were united by a bond that has linked our merchant mariners for more than two centuries—a love of the sea. As their ship battled the storm, they were no doubt working as they lived—together, as one crew. This tragedy also reminds us that most of the goods and products we rely on every day still move by sea. As Americans, our economic prosperity and quality of life depend upon men and women who serve aboard ships like the *El Faro*.

I thank everyone across our government and in the private sector who worked so tirelessly, on the sea and in the air, day after day, in the massive search for survivors. The investigation now under way will have the full support of the U.S. government, because the grieving families of the *El Faro* deserve answers and because we have to do everything in our power to ensure the safety of our people, including those who work at sea. Today, 28 American families—from Florida to Maine—and five Polish families are heart-broken. May they be comforted, in some small way, in knowing that they have the love and support of their neighbors, the merchant mariner community and the American people. May God bless the men and women of the *El Faro*. May He comfort their families. And may He watch over and protect all those who serve at sea on behalf of us all.

The search was over. The grief was only beginning.

As if on cue, Joaquin, which had continued north into the middle of the Atlantic and was now about six hundred miles west of the Azores, weakened and drifted out to sea. Wednesday at

11:00 P.M., a forecaster at the National Hurricane Center typed in the final entry related to the storm. After giving the latitude and longitude, wind speeds, and expected track, the forecaster wrote, "Joaquin has become a post-tropical cyclone. This is the last advisory."

Chapter 14
PORT-DE-PAIX

The wide-open sky above Great Inagua was still and empty, the air limp in defeat. The island had been shaken like a rug, raked over by the winds, soaked by the rains. Debris littered the ground. Downed palm trees and chunks of roofing clogged roads already submerged under water. Houses were flayed, stripped of their siding and shingles, reduced to beams and studs. Yet previous storms had taken much more—lives, not just property—and so the Bahamians persisted, as they had for centuries, clinging hard to this rock like barnacles no hurricane could dislodge. The storm was gone, and now it was time to clean up.

All day Friday, the day after they'd been rescued, and into Saturday, the crew of the *Minouche* bivouacked at the motel, recovering—psychologically as well as physically—from the sinking of their ship. They were two Filipinos, a Dominican, and nine Haitians who sailed under the Bolivian flag. No nation came calling for them or launched ships to find them. They were not on the front page of any newspaper. There were no hearings planned to investigate the ordeal of these itinerant mariners, no cries of "never again." It wasn't that no one cared about them. The *Minouche*'s owner, Milfort Sanon, was concerned about his crew. He knew of the rescue and had informed the families back in Haiti. But he knew little else. These men had embarked on a life at sea, and the sea is unforgiving, especially to sailors who lack even the threadbare protection of a nation's laws. The men of the *Minouche* sailed under a flag of convenience—the flag of a landlocked country—and they were simply not on the world's radar.

But there is one force that mariners can rely on with some degree of certainty to come to their aid: other sailors. Merchant ships, fishing trawlers, the coast guards of most any nation—all are duty-bound to help. Anyone sailing on the open ocean knows how lonely and frightening the vast space around them is. Most sailors would consider it incomprehensible to abandon another sailor to that primal fear. The captains of the *Falcon Arrow* and the *Cronus Leader* both rerouted their ships to help the *Minouche,* at significant expense and some risk. U.S. Coast Guardsmen launched into a ferocious storm to save sailors they knew were not countrymen. A sailor's obligation to protect endangered sailors is formalized in the language of the International Convention for the Safety of Life at Sea (SOLAS), a global treaty governing ship conduct. Shipmasters are required "a) to render assistance to any person found at sea in danger of being lost; b) to proceed with all possible speed to rescue of persons in distress, if informed of their need of assistance." But long before it was enshrined in any official document, rushing to the aid of a foundering ship had been ingrained in maritime culture, a basic tenet of life at sea.

It's possible to find stories of callous captains steaming by ships in distress, not willing to lose time and money to help others. In 2009, for example, the Singapore-flagged freighter *Alam Pintar* slammed into the English fishing vessel *Etoile des Ondes* near the English Channel and didn't stop. More than a dozen ships were in the vicinity, the majority of which failed to respond after a French rescue center sent out a Mayday alert. Luckily three ships did divert course and rescued all but one of the fishing vessel's crew. When contacted later, the captains of the ships that did not respond pleaded ignorance or said they were waiting for direction from authorities. The *Alam Pintar*'s master claimed he thought the fishing vessel was still afloat and unscathed. But after an inquiry, British investigators found that the crew had altered information on the ship's electronics to hide the collision.

Yet incidents like the *Alam Pintar* are so shocking because they are rare. The rescue of the *Minouche* is the norm, a tale that robustly affirms the decency of captains and the men and women who sail under their command.

On Great Inagua, the phone lines and cellphone towers remained down for days. No passenger planes or boats were leaving the island. Gelera and his crew could not reach the ship's agent in Miami or their families in Haiti, Guatemala, and the Philippines. The U.S. Coast Guard had done its job and saved their lives, and was no longer responsible for them. Gelera knew this, and it never even crossed his mind to find his way back to the Coast Guard base and ask if he could use the U.S. military's fancy satellite phones and special communications equipment to get a message out. The international brotherhood of mariners aside, Gelera's years at sea had conditioned him to rely on no one but himself.

Except—sailors are never alone. There are always other sailors.

On Friday morning, some of the men wandered down to a local dock. They asked around until they found a decently equipped wooden boat about fifty feet long with a single one-hundred-horsepower engine. Gelera negotiated with the captain for passage to Port-de-Paix, across roughly ninety miles of open water. This involved an agreement to buy four drums of diesel. Back at the motel, the crew had a meeting to discuss taking the boat to Haiti. Three crew members refused. The sinking of the *Minouche* was too fresh a trauma. *Captain,* they protested, *this wooden boat has no life jackets, no life raft!*

"I will pray for us, that is your life jacket," their captain told them, emboldened by his belief that a miracle had saved them all. These men may have trusted their captain, but they were struggling with the limits of how much. "Also," he told them, "after a hurricane, the seas are calm, calm, calm."

By the next day, Saturday, October 3, 2015, with Hurricane Joaquin well and truly past, the three holdouts had changed their minds. The men of the *Minouche* were finally bound for Haiti.

The fuel arrived on the boat that afternoon around four o'clock, and they motored out from Great Inagua at 9:00 P.M. Just in time: The crew of the *Minouche* had run out of money to pay for food and lodging. The sea was tranquil, just as the captain had promised, and the sky was clear, pocked with stars. Gelera sat out on the deck and kept a watchful eye on his men, worried that someone might suffer a traumatic break and jump in the water. He sat, slept, and prayed. *Thanks to God. Thanks to God, we are finally going back home.* At 6:00 A.M., they arrived in Port-de-Paix. The boat anchored offshore and two rowboats came to meet them and row the men ashore. The wharf was still mostly quiet, but as the men hoisted themselves up onto the dock, the curious came out to see who they were. First a security guard, then some of the stevedores on shift. Like a ripple in a pond, the news spread.

As Gelera made his way to the passport office to register himself and his crew, people started following. Soon they were laughing, then singing, then dancing. *They are still alive!* the crowd shouted. *They are still alive!*

The word on the docks in Port-de-Paix had been that the sailors were all dead, even though Sanon had repeatedly told the families that they had been rescued. But when no one heard from the men, and they couldn't be reached, no one believed him. Now, here they were, standing before the crowd as if risen from the depths of the sea. Some of the sailors' families were summoned. They rushed to the docks with tears in their eyes to embrace their husbands, fathers, brothers, and sons. The other sailors just wanted to get to their homes.

Before letting his crew disband, Gelera gathered everyone together. He hugged them, one at a time. "This is your second life," he told them. "Use this life properly. Believe in God, even if you don't see Him."

The families began planning an impromptu party to celebrate. *Please, Captain, can you stay?* the crew members asked. But Gelera was a man with a family, too, and a driving desire to see them. Now, he was finally able to talk to his wife on a cheap cellphone he had bought in Great Inagua to let her know he was okay.

Ana Santa Maria de Gelera had been trying to find out about her husband for two days. She was in Austria at the time, helping their youngest daughter move there to attend university. One of Renelo's friends in Miami sent her an email letting her know what had happened to the *Minouche*. She immediately tried calling the Coast Guard on Great Inagua as well as the ship's agent, but couldn't get any information. Their son, back in Guatemala, was able to calm her down a little by saying, "Mom, my dad is a wise man and he should have managed to survive, he is fine you will see."

And now she would see, as he was headed to Port-au-Prince to catch a flight back to Guatemala.

Back in Jacksonville and elsewhere, families began returning to homes far emptier than they had been before. Almost immediately, the gears of the U.S. government started turning. Within days, the National Transportation Safety Board (NTSB) announced it was opening an investigation and dispatched a "go team" to Jacksonville.

The Coast Guard also set up a Marine Board of Investigation to examine the cause of the accident. The two agencies ultimately joined forces, but would each write separate reports. The broad goal of the joint NTSB–Coast Guard hearings was to determine what had happened aboard *El Faro* that had led to its sinking. More specifically, the investigators were keen to discover whether flawed decision-making by the officers or neglectful maintenance of the ship by TOTE Maritime—either of which could theoretically rise to the level of criminal negligence—had contributed to the disaster. Finally, the government wanted to do everything possible to identify mistakes that could be avoided in the future.

One of the first orders of business was to find the ship, and with it the voyage data recorder. This was the device that recorded all conversations on the bridge for a minimum of twelve hours, and often much longer. The VDR was contained in a safety-orange canister attached to a mast on the roof of the bridge.

The search for *El Faro* on the surface of the sea may have ended, but a new search below the waves was about to begin.

On October 19, 2015, the U.S. Navy ship *Apache,* a 226-foot ocean tug, set sail from Virginia to find the wreck of *El Faro.* The *Apache* had side-scanning sonar and a deep-sea submersible called CURV-21, an eight-foot-long, 6,400-pound remotely operated machine that could descend to twenty thousand feet. The ship planned to sail a thirteen-segment search pattern fanning out from *El Faro*'s last known coordinates. On October 31, while sailing only the fifth pattern, the sonar towed by the *Apache* pinged on a long whitish image at a depth of about fifteen thousand feet, nearly three miles down. With the sonar's lights on, the ovoid shape clearly resembled a ship and even cast a shadow on the seafloor. The *Apache*'s crew sent down the CURV-21, tethered to the mother ship by a fiber-optic cable. The submersible beamed back full-color pictures. And there the ship was, resting on the seabed, nearly intact. The words *"El Faro"* were distinct on the stern, a hole punched between the words, possibly by the force of the impact as the sinking ship slammed into the ocean floor. A line dangled over the side, an indication, perhaps, of an effort to board a lifeboat. The images were disconcertingly clear. The ship was alone, upright, ghostly. The empty main deck looked as if someone might walk across it any minute.

However, the main target of the *Apache*'s mission—the mast with the VDR attached—was missing. The entire first two decks of the superstructure, including the bridge, had sheared off. Somehow, the waves and wind had ripped a metal structure the size of a house from the upper half of the ship.

Nearly two weeks later, on November 11, the *Apache*'s submersible located the detached bridge lying flat in the sand, as if plucked by a giant hand and gently placed upright half a mile from the rest of the ship. Still, though, no VDR. The mast and the orange canister containing the recorder were nowhere to be found.

The NTSB announced this development on November 16, acknowledging the setback but vowing to carry on.

"Over the years we've completed many investigations without the aid of recorders and other investigative tools," said Christopher Hart, the agency's chairman. "While it is disappointing that the voyage data recorder was not located, we are hopeful that we'll be able to determine the probable cause of this tragedy and the factors that may have contributed to it."

For five more days the *Apache* kept searching for the data recorder, while also continuing to take video of *El Faro*. They did not find the VDR. Finally, the ship gave up and headed back. The mission was over. There were no plans to search again.

But within months the NTSB officials had a change of heart.

The VDR was critical, and the conditions for finding it remained feasible. The CURV-21's images from that depth were remarkably clear and showed an unobstructed seafloor that resembled a desert. In April 2016, the agency announced another search mission. This time, the NTSB contracted the research vessel *Atlantis,* a Navy-owned ship run by the Woods Hole Oceanographic Institution on Cape Cod in Massachusetts. The eighteen-year-old, 274-foot ship also carried a deep-sea submersible, named Sentry. On April 26, Sentry located *El Faro*'s mast, still attached to a steel beam, along with the basketball-sized VDR. Now they had to figure out how to retrieve it. Unlike the CURV-21, Sentry did not have manipulating arms that could pick up the VDR, mast, and steel beam, or cut the VDR free. That would have to be done in a separate operation and expedition.

More than three months later, the *Apache* sailed again to the site, and on August 8, 2016, the CURV-21 descended nearly three miles to where the VDR lay on the lightless black of the ocean floor. Using tools designed for the mission, it clipped the device from the mast and brought it to the surface.

Finally, investigators, TOTE officials, the Coast Guard—and, most important, the families of the lost—stood a chance of getting some answers. The NTSB recovered twenty-six hours of information from the ship, including navigational data, wind speeds, and radar images, as well as all conversations on the bridge. The recordings started at 5:37 A.M., Wednesday, September 30, 2015, and ended at 7:39 A.M., Thursday, October 1. The transcription would take months to complete and eventually fill nearly five hundred pages.

Those pages contained the story of *El Faro*'s final moments.

Chapter 15

"I'M NOT LEAVING YOU"

EL FARO, MORNING OF THURSDAY, OCTOBER 1, 2015

The growing danger of *El Faro*'s situation after it lost power quickly became apparent to Captain Michael Davidson. He had set a course to sail south of Hurricane Joaquin, but because of his reliance on faulty information, and his refusal to heed the warnings of his officers, ended up sailing his ship just north of the eye. Water flooding the holds had given the ship a dangerous starboard list. As a countermeasure, the captain turned the bow north in an effort to stabilize the vessel. Then the engine stopped. Now *El Faro* was drifting helplessly while the counterclockwise winds from the hurricane pushed the ship south, toward the storm's center.

At 7:12 A.M., Davidson ordered his second mate, Danielle

Randolph, to send the distress message she had composed on the bridge's computer. This wasn't Randolph's scheduled shift, but she couldn't sleep and had come up to the bridge to see if she could help.

"All right, Second Mate, send that message," Davidson said. "Push the SSAS [ship security alert system] button."

"Roger," she said.

"Well, all hell's going to break loose with the messaging and stuff like that," Davidson noted, still somehow concerned about how his decision to sound the distress message might make him look in the eyes of TOTE's brass.

"Distress button's been activated," Randolph announced.

"Wake everybody up! Wake 'em up!" the captain shouted. Then, more calmly, "We're going to be good, we're going to make it right here."

The evidence coming in was to the contrary.

Chief Mate Shultz returned from belowdecks. "I think the water level's rising, Captain," he said as he came in the bridge's door.

"Okay, do you know where it's coming from?" Davidson asked.

Shultz said the engineers had told him they thought a fire main had ruptured, but they weren't sure. Fire mains are the pipes that pump seawater in from outside the ship to suppress fires and to be used as wastewater for flushing toilets. The pipes were thick and connected to an endless supply of water. If one had burst, the flooding would be disastrous. But again, Shultz wasn't sure. The water was too high in the holds below to check the fire main's pressure gauge.

Then the chief engineer called the bridge from the engine room to report that the bilge alarm in Hold 2A, the one next to Hold 3, had gone off, indicating that water was flooding into that hold as well.

Shultz hung up the phone and told the captain he was going

down to inspect. But first they discussed the increasing list. Shultz recommended advising the engineers to open all the holds to suck the water out. "I don't know what else to suggest," he said.

Captain Davidson promptly got on the phone with the engine room. "Can you suck on all of the cargo holds?" he asked. "You think the list is getting worse? Yeah, me, too."

Shultz noted that when he looked down into Hold 3, the water was deep and cars were floating.

"They're subs," he said.

Davidson laughed.

"No RPM," someone said, meaning the engine was still down. "We can't do anything."

Getting rid of the water in the holds was a prerequisite to recovering power. If the water could be removed, the ship would right itself, which would allow the oil to reach the pumps and the engine to function again. But they were adrift in a hurricane and rapidly running out of time.

"So," Davidson asked, "what option do we have?"

"Isolate the fire main," Shultz suggested.

Davidson picked up the phone and called down to the engine room. He got first assistant engineer Keith Griffin, whom he referred to as "First."

"Can you isolate the fire main from down in the engine room?" he asked him. "'Cause that may be the root of the water coming in."

Davidson turned to Shultz. "Is there anything else you want me to tell them?"

Shultz suggested shutting off the fire main going forward.

Davidson got back on the phone with Griffin. "Say, First, on the engine room side, the isolation valve suction fire pump, kindly secure it on your side so there's no free communication from the sea, all right? Thank you."

The captain was keeping calm amid all the pressures that must have been weighing on him, including his duty to keep his crew safe. The last thing he wanted to do was put everyone in lifeboats and launch into the ferocious sea raging outside. The captain and his men were now falling back on training distilled to pure instinct while they focused on getting that engine running.

A later analysis of the ship's systems by the Coast Guard would indicate that there was no evidence the fire main had burst. (The NTSB would leave open the possibility.) Instead, it's likely the downflooding angle of the ship that had exposed its ventilation shafts to the waves was the primary source of the flooding.

"Drifting south-southwest," Randolph updated everyone.

"All right," Davidson acknowledged. He reached for the phone and called down to the engine room again.

"All right, Captain here. Chief there?" he asked. "All right, who's this?" It was Jeffrey Mathias, the TOTE engineer on board to supervise the Polish workers.

"All right," he continued, "what did you see down there when you went and looked through that scuttle? Oka. . . . all righ. . . . all right. It's looking pretty nasty. The downflooding angle? I don't have an answer for ya. Yup. . . . What's it called again? Okay, we'll check that. . . . All right, thank you. . . . Um, no, I mean, we still got reserve buoyancy and stability. . . . All right, we're going to ring the general alarm here and wake everybody up."

After a few more minutes of talking to the engine room, the captain hung up and turned to Shultz.

"Just make a round two-deck, see what you can see. This isn't getting any better," he said. "You all right?"

"Yeah," Shultz answered. "I'm not sure I want to go on second deck. I'll open a door down there and look out." Then he added, "Let me grab someone to come with me."

"Okay," Davidson replied.

Almost exactly one minute later, Davidson made the decision to raise the disaster preparedness level. "I'm going to ring the general alarm," he said on the phone to the engine room. But, he added, it didn't mean they were going to abandon ship yet. "We're going to stay with it," he said.

"Ring it!" he shouted at 7:27 A.M.

The peal of the high-frequency alarm tore through the ship, alerting all crew to gather at their assigned muster stations.

"There ya go," Davidson said.

Just then, the chief mate called the bridge on the UHF radio.

"Go ahead, Mate," Davidson said. "Everybody starboard side. All understood."

In desperation, Shultz wanted everyone on board to move to the starboard side of the ship in an effort to counterbalance the port list.

Davidson issued a series of directives into the UHF to Shultz. He wanted everyone in their immersion suits and mustered. "Get a good head count," he said. The captain had accepted the inevitable: They were getting ready to abandon ship.

The ship's list was growing more acute. Water was sloshing onto the decks.

"All right, I got containers in the water!" Second Mate Randolph yelled, as stacks of shipping containers on deck began sliding into the sea.

"All right, let's go ahead and ring it," the captain declared. "Ring the abandon ship."

The alarm, seven high-frequency pulse tones, knelled through the decks, declaring the ship lost. The crew were in survival mode now.

"Can I get my vest?" Randolph asked.

"Yup," the captain said. "Bring mine up, too, and bring one for [helmsman Frank Hamm]."

"Please!" Hamm yelled.

"Okay, buddy, relax," Davidson said. He sensed the helmsman was panicking. The captain turned back to Randolph. "Go ahead, Second Mate," he told her.

Then, looking out the bridge's windshield, he said firmly, "The bow is down. The bow is down."

"Gotta get my wallet and medicine and stuff," Hamm said, asking permission to go below.

"Go ahead," Davidson said. But Hamm didn't leave.

Chief Mate Shultz was still out mustering the crew. The captain called him on the UHF radio. "Get into your rafts!" the captain ordered. "Throw all your rafts into the water!"

"Everybody, everybody get off! Get off the ship! Stay together!" he shouted into the radio.

At the same time, Hamm was trying to get his attention. "Cap," he said. Then again: "Cap."

"What?" Davidson snapped, finally turning his attention to the helmsman. Whatever he saw, he didn't like. The ship's list was at its most extreme by then, more than forty-five degrees. Waves had already overtaken the bow. The floor was tilted at an angle that Hamm apparently couldn't manage to walk up. He was clinging to the console, trying not to slip.

"Come on," Davidson said. He had secured himself in the bridge, and was reaching out. "Gotta move. We gotta move. You got to get up. You got to snap out of it. And we've got to get out."

"Okay," Hamm replied.

"Come up," Davidson said.

"Okay," Hamm replied. Then, "Help me."

"You got to get to safety!" the captain shouted. "You got to get to safety!"

Somebody yelled for the captain.

"What?" he roared back.

"You all right?" they asked.

"Yep," he said.

The bridge was a cacophony of sounds now, alarms, shouting. The phone was ringing. Someone erupted in "Whoo!"

But Hamm remained stuck in his position, unable to reach

the captain. And Davidson was apparently unable to reach him, the ship's list simply too steep. *El Faro* at this point was sinking beneath the waves, listing to port, air in the hull keeping it just above the water as the starboard side lifted up and threatened to come crashing over, flipping the boat upside down.

"What do you think?" Hamm yelled. "Help me. Help me!"

The captain called out his name. "Don't panic," he said. "Don't panic, don't panic. Work your way up here."

"I can't!" Hamm yelled back.

The captain called his name again.

"Help me," Hamm continued to plead.

"You're okay, come on," Davidson coaxed.

Someone yelled for the captain.

"Where are the life preservers up here?!" the captain shouted over his shoulder.

Then he turned back to Hamm.

"Hey," he said loudly. "Don't panic."

The captain spoke to some others about getting their life vests on.

Then Hamm yelled, "I can't!"

"Yes, you can," the captain said, loudly, firmly, as if giving an order.

"My feet are slipping!" Hamm yelled. "Going down!"

"You're not going down," the captain said. "Come on."

"I need a ladder!" Hamm yelled.

"We don't have a ladder," Davidson replied.

"A line!" Hamm said.

"I don't have a line," Davidson responded.

"You gonna leave me," Hamm cried out.

"I'm not leaving you," the captain replied. "Let's go!"

Hamm screamed. Then pleaded, "I need someone to help me!"

"I'm the only one here." Davidson struggled to sound calm amid the mounting horror.

"I can't!" Hamm yelled. "I'm a goner."

"No you're not!" Davidson insisted.

"Just help me," Hamm pleaded again.

The captain couldn't reach him, but he did not abandon him.

"Let's go!" he said loudly.

There was yelling in the background. Then a low rumble. Davidson called out for Hamm one more time.

"It's time to come this way!" Davidson yelled.

Those were the last recorded words from the bridge of *El Faro*. It was 7:39 A.M., six minutes after Chief Petty Officer Chancery in Miami received his email that the ship was having some difficulties, but was not in imminent danger.

The details of what happened next can never be known. The body the Coast Guard discovered suggests there was enough time for some crew members to put on their immersion suits, which would have helped with flotation and protected against hypothermia, although hypothermia would not have been the most pressing concern in 86-degree water. The two forty-three-person fiberglass lifeboats were found smashed up, one half-submerged, the other on the ocean floor. Experts who examined them said it was unlikely either lifeboat was launched by the crew. There were the four twenty-five-man and one six-man inflatable life rafts that the captain had told his chief mate to deploy. But scrambling down the steeply tilted decks of a sinking behemoth like *El Faro* and trying to board inflated life rafts in violent, chaotic seas—all while visibility was blocked by volumes of rain that transformed everything into a dense gray smear—would not have been easy.

The vessel would have been shuddering and rocking violently as it slipped beneath the waves. If the life rafts remained tethered to the ship, they would have been getting tossed around wildly on the surface, almost impossible to rein in and control. If they weren't tethered, they would have simply blown away.

If any of the crew had somehow managed to throw a line or a

ladder over the stern and make their way into a raft in time, what then? They were twenty miles from the eye of a Category 3 hurricane. The sustained winds were 120 miles per hour, strong enough to sweep grown men off the deck of a massive cargo ship, let alone a life raft that was no more than a bouncy house in the waves. Heavy, violent sheets of rain would have pummeled them from above. The wind would have whipped and aerated the surface of the sea into a thick white spume, the air just above the surface dense with a misty spray known as spindrift. In essence, the surface above the water would have been a semi-aqueous state. Coast Guard rescue swimmers deployed in such conditions have recounted extreme difficulty breathing, even with a snorkel. A survivor in an immersion suit, with its limited mobility, would have been severely challenged to keep an air passage open. In these conditions, it is possible to drown while still afloat.

And then there were the waves. Towering thirty-to-forty-foot walls of water with overhanging crests would have loomed over and then crashed down on them. Whether the sailors were in a raft or afloat in an immersion suit, breaking waves of this size would have been powerful enough to hold them underwater long enough to drown. The waves would have also been strong enough to turn any debris in the water—floating cargo containers, chunks of metal ripped from the ship, even cars—into seaborne missiles. This flotsam could come crashing down with enough force to smash life rafts in two and crush any human occupants.

As the ship rolled over and then started to sink, it would have created an air pocket that then filled with water, sucking down everything around it.

Investigators who later examined photos of the sunken ship believe that it capsized on or near the surface, turning completely upside down, but then righting itself before it reached the bottom. When the 883-foot *Titanic* split in two and sank in 1912, the sections are believed to have descended at between 35 and 50 miles per hour. *El Faro,* only slightly smaller, would have sunk at a similar rate, speeding the three miles to the bottom in less than four minutes. The ship's rudder likely hit the bottom first, punching the propeller shaft through the stern, where the ship's name was painted. The bridge section was ripped off, probably during the capsizing, and drifted off like a leaf of paper on a breeze to settle half a mile away. And then it was over. The hull settled heavily into the silt, unlikely to ever move again.

Chapter 16
EIGHT BELLS

The practice of using bells stems from the days of the sailing ships. Sailors couldn't afford to have their own time pieces and relied on the ship's bells to tell time. The ship's boy kept time by using a half-hour glass. Each time the sand ran out, he would turn the glass over and ring the appropriate number of bells. Each ship "watch" is four hours, or eight bells, in length. . . . The tradition of Eight Bells pays respect to deceased mariners and signifies that a sailor's "watch" is over.

—Maine Maritime Academy

In the immediate aftermath of the wreck, there was an absence of answers. As a result, lawsuits began filling the void, as well

as the docket of the federal court in Duval County, Florida. Grief
and anguish were barely subdued beneath dry legal language citing admiralty law. The family of assistant steward Lonnie Jordan
filed a $100 million lawsuit against TOTE a week after the Coast
Guard announced the end of the search. Jeremie Riehm's widow,
Tina, filed a wrongful death suit on the very day the *Apache* set
sail to look for the ship. Two weeks later, able seaman Carey
Hatch's widow, Tracey, filed her suit. And on and on. In brief
after brief, the plaintiffs accused TOTE of negligence in managing, operating, and maintaining *El Faro* and safeguarding her
crew. At their core, the lawsuits all posed the same underlying
question: *How could this have happened?* How, in the twenty-first
century, with all the technology available to us, did a ship run by
a modern American corporation bypass all safeguards, all warnings, and sail into a hurricane?

Regardless of whether a lawsuit was filed, TOTE, which
would receive a $36 million insurance payment for *El Faro*'s loss,
began offering initial $500,000 payments to crew members'
families. Eventually the company settled with all the U.S. families, paying additional, confidential amounts based on specific
information about the individual sailors. Presumably the cold,
impersonal logic of the law was used to estimate potential future
earnings of the deceased, how much he or she was being paid for
the final voyage, the number of dependents left behind. The ac-
tuarial tables and accounting metrics that were deployed could in
no way reflect the magnitude of losses far more profound than
the depth at which *El Faro* rested.

In the aftermath of the tragedy, first assistant engineer Keith
Griffin's widow, Katie, gave birth to twin girls who will never
know their father. The parents of newly minted third assistant
engineer Dylan Meklin sent their twenty-three-year-old boy off
to his very first job after graduating Maine Maritime Academy
that summer. It was supposed to be the beginning of a life, not
the end.

How do you measure the difference in tragedy between a child growing up without a parent or the parent who loses a child? In one way or another, this formula for measuring loss was being calculated thirty-three times, for every sailor who perished on the SS *El Faro*'s final voyage:

ABLE SEAMEN: Jackie Robert Jones, Jr., Larry Davis, Frank Hamm, Carey Hatch, and Jack Jackson.

OILERS: Joe Edward Hargrove, Anthony Shawn Thomas, and German Arturo Solar-Cortes.

QUALIFIED MEMBERS OF THE ENGINE DEPARTMENT: Sylvester Crawford, Jr., and Louis Marko Champa.

GENERAL UTILITY DECK ENGINEERS: James Phillip Porter, Mariette Wright, and Roosevelt Lazarra Clark.

ASSISTANT STEWARD: Lonnie Jordan.

CHIEF COOK: Lashawn Lamonte Rivera.

STEWARD-BAKER: Theodore Earl Quammie.

BOSUN: Roan Ronald Lightfoot.

CHIEF ENGINEER: Richard Pusatere.

FIRST ASSISTANT ENGINEER: Keith Griffin.

SECOND ASSISTANT ENGINEER: Howard John Schoenly.

THIRD ASSISTANT ENGINEERS: Mitchell Kuflik, Michael Lee Holland, and Dylan Meklin.

THE TOTE CHIEF ENGINEER SUPERVISING THE POLISH WORKERS: Jeffrey Mathias.

CHIEF MATE: Steven Shultz.

SECOND MATE: Danielle Randolph.

THIRD MATE: Jeremie Riehm.

POLISH WELDERS: Jan Podgórski, Marcin Nita, Andrzej Roman Truszkowski, Rafal Andrzej Zdobych, and Piotr Marek Krause.

And, finally, SHIPMASTER Michael Christopher Davidson.

For the grave and terrible responsibility of sailing his ship into a hurricane despite repeated warnings, Captain Davidson has much to answer for. Yet it's also true that Davidson spent the last minutes of his life extending his arms to help one of his crew. He didn't abandon helmsman Frank Hamm or leave the bridge. Some of that should be remembered, too.

Relatives of the dead were not the only ones to sue the company. The customers who had paid to have their goods shipped to Puerto Rico also weighed in.

One eddy in the stream of lawsuits was filed by insurance giant Lloyd's of London, representing about a dozen policyholders, against "StormGeo Corp. DBA Applied Weather Technology," the makers of the weather-tracking Bon Voyage System. The lawsuit alleged that StormGeo was negligent and liable because it falsely advertised that their product provided current weather data when, in fact, the information it broadcast was "highly delayed." The suit blamed StormGeo for leading the captain astray.

"The data provided by the BVS 7 product with respect to the center and the projected track of the Hurricane was defective, because the readouts consisted of little more than publically [sic] available data from at least nine to thirteen hours earlier being reproduced and utilized and sold by AWT to its consumers on a highly delayed basis," the lawsuit stated.

StormGeo's lawyers contend the lawsuit confuses weather data and weather forecasts, asserting that the information takes hours to check and add to, and that once it is sent to the ship, the Bon Voyage software uses the data to generate the most current predictions and forecasts. For more urgent updates, the program can be set to deliver NHC information sooner. But that option had not been activated on *El Faro,* the lawyers say. As of this writing, the lawsuit was pending.

Even as lawsuits were filed and testimony heard, *El Faro*'s sister ship *El Yunque* came due for a thorough inspection. *El Yunque* was also a Ponce-class vessel, built by Sun Shipbuilding and launched in 1976, making it one year younger than *El Faro*. The two ships were nearly twins. So *El Yunque* now faced intense scrutiny.

Throughout December 2015, the Coast Guard pored over the ship, discovering deep "wastage," a term used to describe rusting so thorough it affects the steel's integrity. The inspectors wrote, "Several break downs have been observed with fundamental systems that enhance shipboard safety . . . to include damaged and unrepaired CO_2 fire fighting system in cargo spaces, clogged and completely wasted second deck sprinkler system piping, major rudder post seal leaks, and inoperative steering room ventilation. These shortfalls are an indication that a properly implemented preventative and corrective maintenance system, to include adequate documentation, does not exist on board the vessel."

The Coast Guard put a "no sail" order on *El Yunque* until repairs were made. An American Bureau of Shipping surveyor witnessed and validated the repairs, and the order was lifted. But the Coast Guard returned and found more corroded and clogged sprinkler piping. In February 2016, a team of inspectors found ventilator shafts so rusted that a "hammer test"—simply tapping metal with a hammer—punched a hole through the baffle plating.

As the Coast Guard inspectors began talking to the ship's crew about expanding their examination to other ventilation shafts, the Coast Guard's commanding officer in Jacksonville, Captain Jeffrey Dixon, called the senior inspector and ordered him to halt their work because it exceeded the scope of the Coast Guard's audit. (Fifteen months later, Dixon would retire and join TOTE as vice president, marine operations, government and commercial.) The inspector complied with the order—at the time. Later, though, suspecting the corrosion in the other ventilation shafts was significant, he put in a request for additional inspections.

El Yunque wouldn't escape scrutiny for long. In March 2016, she sailed up to Puget Sound, in Washington, to take *El Faro*'s place running the coastal route to Alaska. The Coast Guard targeted her as soon as she came to port. Despite ABS reports verifying that TOTE had repaired the wastage found in Jacksonville, Coast Guard inspectors discovered severe corrosion in other ventilation shafts.

When TOTE realized the extent of the repairs the ship would need to satisfy the inspectors, they gave up. In August 2016, *El Yunque* was hauled to a Texas scrapyard and dismantled piece by piece with an acetylene torch. Inadequate payment for the sins of *El Faro*'s neglect, perhaps, but at least it was one unsafe ship removed from service. Two other ships not owned by TOTE, the *Energy Enterprise* and the *Captain Steven L. Bennet,* were also sent to the scrap heap after the heightened scrutiny following the sinking of *El Faro.*

Coast Guard officials have testified that a review of *El Faro*'s inspection reports didn't reveal the kind or extent of rusting that was found on *El Yunque.* But those inspections had been performed by the American Bureau of Shipping. Given how much the ABS inspectors had missed on *El Yunque,* how certain can anyone be that *El Faro* wasn't suffering the same or similar levels of deterioration? If *El Faro* had been subjected to as thorough a probing as *El Yunque,* would she have been allowed to sail?

Chapter 17
THE MASTER'S JUDGMENT

Jacksonville's Prime F. Osborn III Convention Center, a Greek Revival structure complete with massive Doric columns, is a precocious emblem of a bygone era. It was originally a rail depot meant to celebrate a turn-of-the-century city thriving in timber and shipping. In the years since, the city has never quite lived up to the center's grandiosity. But its somber façade made an appropriate home for the Marine Board of Investigation (MBI) examining *El Faro*'s sinking.

This would be Captain Jason Neubauer's office on and off for the next sixteen months. Neubauer, the chief of the Coast Guard's Office of Investigations and Casualty Analysis in Washington, D.C., had, in a sense, been working on the case since the morning Danielle Randolph sent the ship's distress signal. He had been in his D.C. office at Coast Guard headquarters that morning, when he saw a critical-incident communication flash on his screen. Something about a deep-draft vessel having some issues in a hurricane.

"The word was we were having trouble contacting it," recalled Neubauer, who keeps his blond hair trimmed short on the sides and speaks in the carefully measured tones of someone who values precision. "I started thinking the worst."

That, after all, was his job. He emailed the National Command Center and asked if anyone was looking at this notification from a casualty perspective. "I think people didn't believe we could lose a 790-foot vessel," he surmised. But he remembers thinking, If this ship sinks, we will have one of the biggest casualties in the history of Coast Guard investigations. In fact, it was the largest marine disaster involving a U.S. ship in thirty-nine years, since the ferryboat *George Prince* collided with the Norwegian freighter *Frosta* on the Mississippi River in 1976, killing seventy-eight people.

On October 8, 2015, after the search and rescue mission for *El Faro* concluded, the Coast Guard convened the MBI to look into the accident. The board comprised a joint task force of Coast Guard advisers, investigators, and technical experts from throughout the service.

To avoid duplication of effort, the National Transportation Safety Board had its own investigation staff, led by Thomas Roth-Roffy, participate in the MBI hearing sessions. Roth-Roffy was a sailor, a graduate of the U.S. Merchant Marine Academy. He had served as an engineer and chief engineer for the Military Sealift Command before going back to graduate school and then joining the NTSB.

From a raised dais, the board interviewed seventy-six witnesses—TOTE executives, sailors, engineers, Coast Guardsmen—to try to learn everything it could about the company, the men and women on board, and the jobs they did. On a lower dais, "parties in interest" were allowed to sit and ask questions. These included TOTE Inc. and its subsidiaries; the American

Bureau of Shipping, which had performed inspections of *El Faro;* Herbert Engineering Corporation, the firm that worked on the ship's conversions in 2005; and lawyers for Theresa Davidson, Captain Davidson's widow. The stated reason for this arrangement was so that investigators would have immediate, in-person access to their knowledge of the company, vessel, and events. But the fact remained that these were people and companies that could be held accountable for negligence or wrongdoing. Having them there served a dual purpose of allowing them to defend themselves and their actions if necessary.

The board met in Jacksonville three times for about three weeks each time: one week for preparation and two weeks of testimony in the convention center. Many family members of the *El Faro* crew attended all three sessions.

There were plenty of tense moments. Roth-Roffy was pointed in his questioning of TOTE executives. When Philip Morrell, TOTE's vice president of marine operations, commercial, testified on February 16, 2016, that the forty-year-old Ponce-class steamships *El Faro* and *El Yunque* were as reliable as newer ships, Roth-Roffy seemed incredulous. "In terms of on-time performance, level of maintenance required?" Roth-Roffy asked, clearly skeptical. Then he added, "Can you speak to why Tote Maritime would maintain these vessels in service, forty years of age?" Ships half that age are routinely retired for scrap, he noted. "Do you have insight on the decision to continue to operate these vessels?"

"No," Morrell abruptly replied.

Then on May 26, 2016, Roth-Roffy grilled TOTE Inc.'s executive vice president, Peter Keller.

"Now, sir, many would argue and few would dispute the loss of the ship *El Faro* and its cargo and most importantly the loss of thirty-three souls aboard the *El Faro* represents a colossal failure in the management of the companies responsible for the safe operation of the *El Faro*," Roth-Roffy stated. "And, sir, you have no doubt thought long and hard about the nature of the management failures that led to the loss of the *El Faro*'s crew. Could you please share with this board your thoughts about the nature of the management failures that led to the loss of the *El Faro*?"

Keller was clearly taken aback.

"I think this tragic loss is all about an accident, and I look to this board as well as the NTSB to try to define what those elements may or may not have been," he smoothly answered. "I, for one, with fifty-one years of experience in transportation, cannot come up with a rational answer. I do not see anything that has come out of this hearing or anything else that I've ever seen that would talk about a cause. Certainly as management we look for that. We look for what NTSB and this board may come up with. Because we think it will be important. At this point in time I for one cannot identify any failure that would have led to that tragic event."

The board would find plenty of failures before it was through.

But not without some resistance from TOTE, which objected to Roth-Roffy's handling of the case. TOTE's lawyers complained to him and his superiors that his line of questioning sounded more like a conclusion than an inquiry, at a time when the board hadn't finished its work. Rather than let the episode turn into a distraction, Roth-Roffy said he thought it would be better to just apologize and continue on. He did so at the next hearing, saying he was sorry if his questioning was misinterpreted.

Then, a short time later, Roth-Roffy announced he was retiring to take a job as the chief engineer on SUNY Maritime College's training ship, the TS *Empire State VI*. This led to speculation that pressure from TOTE had led to his departure. Both the NTSB and Roth-Roffy said this wasn't true, explaining that he had actually accepted the job before the hearings began. "It was time to go back to sea," he said. Still, several *El Faro* family members were sorry to see him go. To them, he had been a staunch advocate willing to ask hard questions.

The NTSB replaced Roth-Roffy with Brian Young, a marine engineer who'd graduated from SUNY Maritime College, and the Marine Board of Investigation soldiered on.

The board heard from widows, parents, neighbors, and friends of the crew, and to a large degree felt they knew the families and even the lost sailors intimately. That gave their job an extra burden.

Neubauer, for one, felt completely accountable to the families. He knew the board had to do a thorough assessment. Often this meant tracking down rumors family members had heard, such as the one that *El Faro* was sailing with only one boiler operating (false), or that the ship's anemometer was not working (true), because they fueled confusion and anger.

Finally, on February 6, 2017, sixteen months after first convening, the board held its final hearing. Before the first witness was called, family members of *El Faro*'s crew placed black ribbons on the first thirty-three seats in the room, and Captain Neubauer opened with thirty-three seconds of silence for the drowned sailors. Captain Davidson's widow, Theresa, wrote a statement that her lawyer read to the board.

> My daughters and I know the pain you feel, as do Michael's siblings and parents. . . . Crew members, both licensed and unlicensed, who sailed with Michael described him as meticulous, concerned for safety, caring for the welfare of his crew and a true professional. But Michael was much more than a ship's captain. He was a brother, a terrific father to two daughters and an amazing husband. And if you really want to know who Michael was at his core, you only need to read the last pages of the VDR. He willingly gave up the opportunity to fight for his own survival because he refused to

leave a crew member behind. Some were surprised Michael made that choice. I was not.

Now it was time for the Coast Guard and the NTSB to retire to their respective offices, digest the volumes of data and testimony, and come to some conclusions independent of each other.

On October 1, 2017, the two-year anniversary of the accident, Neubauer stood at a podium in a small conference room inside the Coast Guard's Jacksonville offices to announce the release of the MBI's final report. It was a gray, wet Sunday, and a handful of mostly local reporters were there; others watched on a live Internet transmission. The office building was not far from the Blount Island terminal from where *El Faro* last set sail. Just down the street, under the Dames Point Bridge, was a memorial park dedicated to the crew of *El Faro*.

"The MBI determined the primary initiating event for this tragedy was the vessel's close proximity to Hurricane Joaquin," Neubauer said, standing in a formal dark uniform. He confirmed:

> The master was ultimately responsible for the vessel, the crew and its safe navigation. The master misjudged the path of Hurricane Joaquin and overestimated the vessel's heavy weather survivability, while also failing to take adequate precautions to monitor and prepare for heavy weather. During critical periods of navigation, where watch standers were looking to the master for his guidance and expertise, he failed to understand the severity of the situation, even when the watchstanders warned him that the hurricane was intensifying and that *El Faro*'s projected closest point of approach was decreasing.

Later, in response to a reporter's question, Neubauer acknowledged that "if the master had survived, we would have pursued a negligence complaint against his merchant marine credentials."

But the Coast Guard was not going to let Davidson take the fall alone. The MBI found that three other parties shared some responsibility, if not directly for the accident, then at least for the poor condition of the ship, which led to its vulnerability to engine failure and downflooding. The first was the American Bureau of Shipping, which had been responsible for inspecting the ship. The second was the Coast Guard itself, which was responsible for overseeing the ABS inspections and conducting followup inspections of its own. And the third was, of course, TOTE Inc. In the report, TOTE is held accountable for every step of the captain's negligence and misjudgment:

> TOTE did not provide the tools and protocols for accurate weather observation.
>
> TOTE and the Master did not adequately identify the risk of heavy weather when preparing, evaluating, and approving the voyage plan prior to departure on the accident voyage.
>
> TOTE did not provide adequate support and oversight to the crew of *EL FARO* during the accident voyage.

And on and on. The Coast Guard emphatically rejected TOTE's position that once its ships left port, they were on their own. There was no reason someone shoreside couldn't be monitoring the ships and the weather and communicate with crews to warn them when they needed to change course to avoid dangerous situations. Many shipping companies had this system in place. "TOTE did not ensure the safety of marine operations and failed to provide shoreside nautical operations support to its vessels," the report concluded. Then it recommended that the Coast Guard create a regulation that oceangoing commercial vessels periodically transmit electronic records back to shore—including significant weather and route changes.

Additionally, the report strongly criticized the company's lack of maintenance, training, and preparation, which in this case had had fatal consequences. The crew members had not been aware of the ship's offset oil sump pump, which made the ship vulnerable to engine failure during a port list. If the crew had known about this vulnerability, they may have been able to avoid the problem in the first place. "Complacency" and "lack of training" led to the crew's failure to assess whether the ship's watertight integrity had been compromised. Shoddy maintenance allowed the wastage in *El Yunque*'s ventilation shafts and likely in *El Faro*'s as well, which would have contributed to the down-angle flooding of the holds.

In that particular, the board found the Coast Guard itself negligent. The Coast Guard's marine inspectors needed to do a better job supervising the ABS's surveyors to make sure they were being rigorous in their inspections.

But the vast majority of that criticism was just a shaming exercise. The board found only four official violations, including evidence that officers did not get the required amount of rest in previous trips; evidence TOTE had not conducted safety drills with the Polish welding crew; that TOTE had not notified the ABS or Coast Guard of changes to lifesaving equipment—replacing clutches on the lifeboat davits—prior to the September 29, 2015, departure from Jacksonville; and that TOTE failed to notify the Coast Guard or the American Bureau of Shipping about repairs to the boiler's superheating piping system. None of these contributed directly to the ship's sinking.

The most damning errors—such as using old-fashioned lifeboats and having no GPS in the EPIRB—were legally permissible.

The board made a series of thirty-three safety recommendations to the commandant of the U.S. Coast Guard. Some were highly technical. Others stood out. The board recommended re-moving the grandfather clause protections allowing older ships to continue using the open-top lifeboats and EPIRBs that are not encoded with GPS. "Even though it's allowed, it's not right and it's not fair to U.S. mariners," Neubauer said. The board also asked the Coast Guard to come up with a plan to extract a DNA sample from a drowned sailor if the body couldn't be recovered, and to test the type of data buoys used to mark the dead *El Faro* sailor and replace them if necessary.

In addition, the report recommended bilge alarms for the holds, indicators on the bridge for all watertight enclosures, closed-circuit TV monitors in all stowage areas, personal locat-ing beacons for all personal flotation devices on commercial ves-sels, and a mechanism to anonymously report safety warnings from ships to either the Coast Guard or the company's desig-nated person ashore.

So, despite TOTE executive Peter Keller's pronouncement that he couldn't see any reason for this tragic event, many factors contributed to it. This is all the more remarkable given that on the same day, the same storm was battering another ship less than two hundred miles away. The *Minouche* was much smaller and far less sophisticated. It didn't have color-coded weather-tracking software. It wasn't part of a large corporation. But maybe the lack of technology meant the captain had to be in tune with his ship and the weather, and be alert to anything, *anything*, that was out of place. In the case of the *Minouche*, Captain Gelera's quick responses to everything from mustering the crew to sending the distress signal helped them all survive.

But when all the accounting is done, all the assignable blame assigned, there's one thing no government can force into compli-ance: the weather. More and more extreme weather events are occurring as our planet warms. Global average temperatures are increasing, ice sheets are melting, the patterns of precipitation are changing, and sea levels are rising. Some climate phenomena are understood—greater moisture in the atmosphere leads to heavier rainfalls. Some are not. Hurricanes have always defied prediction; what ignites them is still unknown. But climate scientists expect the strongest storms to get stronger, citing the Clausius-Clapeyron equation, which indicates air moisture will increase roughly 7 percent for each degree Celsius the sea surface temperature rises. The strongest hurricanes on record have all occurred recently, and that is almost certainly not a coincidence.

Less is known about how changing weather will affect the traditional paths of hurricanes and the duration of the traditional hurricane season. But the influence of climate change on cyclones should not be underestimated. The warmth of the water that Joaquin passed over, for instance, clearly allowed the storm's strength to rapidly intensify.

This hurricane stood out for other reasons as well. "Joaquin's formation is notable in that the cyclone did not have tropical origins, which is rare for a major hurricane," hurricane specialist Chris Berg wrote in the National Hurricane Center's analysis of the storm. Additionally it "was the strongest October hurricane known to have affected the Bahamas since 1866 and the strongest Atlantic hurricane of non-tropical origin in the satellite era."

It formed in an unusual place, took an unusual path—south instead of west, which made it difficult to predict—and passed over waters that were a record high temperature for that time of year. Yet it was just one in an era of remarkable storms. Twenty days after Joaquin sank *El Faro*, Hurricane Patricia formed in the Pacific off southern Mexico's west coast. As it moved over "anomalously warm waters," according to the NHC, it quickly bloomed into the strongest hurricane on record, with peak winds of 215 miles per hour. The storm weakened at sea before making landfall as a Category 4 hurricane.

Unusual is the new usual, it seems, and records are being broken all the time now.

In 2004, for the first time ever, a hurricane originated in the South Atlantic, where wind shears typically blow them apart before they have a chance to form. The storm stumped the World Meteorological Organization, which maintains a list of hurricane names for every region of the world—except the South Atlantic. That year went on to become the fifth most active hurricane season on record, as measured by accumulated cyclone energy. The next year, 2005, was the second most active hurricane season on record. In fact, five of the top ten most active seasons have occurred since 1998.

The 2017 season brought us Hurricane Harvey in August, a Category 3 storm that flooded Houston, Texas, with forty inches of rain, making it the wettest tropical cyclone on record in the United States. In late August and September, Irma, the strongest Atlantic hurricane on record, made landfall in the Caribbean and Florida Keys, with maximum sustained winds of 183 miles per hour. In late September and early October, another Category 5 hurricane, Maria, with 175-mile-per-hour winds, devastated several Caribbean islands and knocked out the power for the entire island of Puerto Rico. Then, in late October, Hurricane Ophelia, which formed outside the tropical belt and took an extremely rare track west to east toward Ireland, became the most easterly hurricane to ever form in the Atlantic.

Even taking into account historical fluctuations in hurricane intensity and frequency, which complicates the science of predicting future trends, climate scientists can state with confidence that stronger storms are on the horizon.

"Future projections based on theory and high-resolution dynamical models consistently indicate that greenhouse warming will cause the globally averaged intensity of tropical cyclones to shift towards stronger storms, with intensity increases of 2–11% by 2100," stated an article by ten international climate scientists in the scientific journal *Nature Geoscience*. "Existing modelling studies also consistently project decreases in the globally averaged frequency of tropical cyclones, by 6–34%."

Translation: While the number of stronger storms is likely to increase, the overall number of storms may go down. Still, "balanced against this, higher resolution modelling studies typically project substantial increases in the frequency of the most intense cyclones, and increases of the order of 20% in the precipitation rate within 100 km of the storm centre."

Yet, in the United States, powerful politicians in the Republican Party—congressmen, senators, and governors—refuse to acknowledge the threats of climate change or act in any way to lower carbon emissions, the one thing humans could do to try to offset our contribution to rising global temperatures. They have not engaged in a dialogue about what we should or can do, nor have they made much of an attempt to explain the evidence on which they base their denial. For more than eight years, congressional committees in the House of Representatives that could deal with the issue have routinely prevented federal action on climate change. Throughout his two terms in office, Rick Scott, the Republican governor of Florida—the state most susceptible to rising sea levels as well as hurricanes—prohibited the terms "climate change" and "global warming" from even being *used* by state agencies.

Every passing minute of their inaction increases their responsibility for the safety of first responders and others affected by hurricanes, historic floods, and wildfires.

Storms and shipwrecks have never deterred sailors from going to sea. If that were the case, we would have run out of coffee and tea long ago. Lost ships are a reality of global commerce. Woven into the grief of the families of the *El Faro* crew is a strain of acceptance; the sea sometimes takes sailors. It's a reality—a universal law—and understanding this can help some to make peace with the loss.

Not everyone, of course. Able seaman Jack Jackson's brother, Glen, with his law enforcement background, pored through the reports and testimony trying to figure out how this "accident" could have happened. Camden, New Jersey, native Tina Riehm still grapples with the suddenness of the loss—how someone could be here one minute and then disappear beneath the waves the next. Compounding her family's grief was when the presents, like a cruel joke, began arriving. Every fall, when Jeremie was away at sea, he would start ordering Christmas presents. They'd arrive months in advance, with a strict policy that none could be opened until he returned, if he was lucky enough to be off for Christmas that year. They were piling up by the door as the holidays approached.

But for those from seafaring families, there is often a solemn understanding that the specter of catastrophe is simply part of this life. "We've got the old saltwater in our veins," said Laurie Bobillot, Danielle Randolph's mother. She understood that the sea is an unpredictable force that can turn on you in an instant. "As beautiful as the ocean is, things develop quickly," Bobillot said. "Storms develop quickly." She holds no ill will against the company. "Though I do feel TOTE could have done more, as in better weather updates and knowing the location of the ship," she said, "they've been good to us."

Claudia Shultz is plenty angry about the mistakes that cost her husband's life. Her lingering regret is that neither Jeremie Riehm nor Danielle Randolph woke the chief mate up after calling the captain to air their concerns. "I know if they had called Steve, he would have knocked on the captain's door and said, 'Look, Captain, we have to talk about this.'"

But that didn't happen, and Claudia, from a family of sailors, understood her husband. "If they had survived, I guarantee you, 100 percent, Steve would be right back out there on a ship."

Carla Newkirk knows this as well. Her father, Larry Davis, was a second-generation sailor in Jacksonville. His brothers and his father were all merchant seamen. Newkirk herself works for a shipbuilding company.

"We all knew that anything could happen out there," Newkirk

said. "There might be a time when someone you love doesn't come home." In this tradition, focusing on the assignation of blame is seen as not healthy. So she is not angry at the captain. "He made the decision he made. It ended tragically. Do I wish he had made different decisions? Yeah, but we can't change what happened."

On a bleak April day in 2017, Newkirk and her oldest son, Tyler, made the ten-hour drive from Jacksonville to the Seafarers International Union's Harry Lundeberg School of Seamanship, in Piney Point, Maryland. After trying a couple of jobs following high school, including working at a warehouse for nine dollars an hour, Tyler had made his choice. He wanted to go to sea, like his grandfather and his great-grandfather before him. He was excited to start his career, although he wasn't sure what he wanted to do yet—work the bridge, the decks, or the engine room. He was pretty sure he did not want to be in the galley. It was an emotional car ride for Newkirk. As a mom, she was proud of Tyler. He was an adult now. But this was her baby and he was leaving home, so there was some sadness mixed in there, too. And then there was the legacy of her father, who they called Papa, hovering over them.

The morning after they arrived in Piney Point, they went to the school's *El Faro* memorial, a lighthouse replica with a brass plaque and a ship's bell mounted on it, set on a circular walkway of red bricks. Thirty-three of the bricks were stenciled with the names of the *El Faro* crew. They found Larry Davis's brick. It had rained the night before. The sky was still gray and overcast, which lent the memorial, which overlooks the St. Marys River, a steel-hued beauty. It was so pretty, in fact, that Newkirk was either going to have to leave or cry.

"We looked at all of that, and talked about Papa, and how proud Papa would be of him right now," she recalled. "I said, 'You know, we're all going to miss you, and your Papa is looking down right now and he's very proud you're following in his footsteps.'"

The wreck of *El Faro* had not scared off this family. A life on the sea was a good life as far as they were concerned. And losing your life at sea was a proud way to go.

About a year before his final voyage, Larry Davis had taken some time off to tend to his older brother, who was dying of cancer. His brother's illness was hard on Davis. So was being on land for an extended time. He tended to get restless after a few months. His brother died six months before Davis shipped out on his last voyage.

"I would never have wanted to see my dad die of cancer in a bed," Newkirk said. "For him, his passion was to be on the ocean. I'm at peace with it. That's where his journey needed to end."

EPILOGUE

The clouds over Mississippi are a dark and mottled gunmetal, the feathery fringes of Hurricane Michael churning in the Gulf of Mexico to the east. Two years after flying a C-130 to the U.S. Coast Guard base in Great Inagua, Bahamas, I was back on a government runway, this time in Biloxi, getting ready to board another C-130, specifically an industrial gray U.S. Air Force Reserve WC-130J, modified with weather-reading instruments and extra fuel capacity.

This book has been done for well over a year now, but I still look for opportunities to write about hurricanes. In this case I've been offered a trip on a hurricane hunter, one of ten in the Reserve's 53rd Weather Reconnaissance Squadron out of Keesler Air Force Base. I'm wearing a long-sleeved shirt, jacket, and hat despite the southern heat. I've been warned that it gets cold inside the stripped-down fuselage at 10,000 feet. It's

October 10, 2018, and the itinerary is to fly into the eye of the storm.

At this point I'm dangerously close to being obsessed with tropical cyclones. Not only have I tried to educate myself about the science behind hurricane formation, I've endlessly read and re-read my notes from the captain and crew of the *Minouche*, trying to understand what the experience of surviving one was like for them. And I'm still haunted by the transcript containing the last words of the sailors from the bridge of *El Faro*. It's an occupational hazard for a writer—to conjure through imagination what you are recounting until your mind can't let go. But the more I learned about hurricanes, the more they defied my imagination—their size, wind energy, water volume, the immensity of the climatological forces at work. It just follows that I wanted to close that gap in my knowledge by getting as close to one as possible.

One opportunity presented itself while I was writing this book. In September 2017, as Hurricane Irma barreled toward the city where I live, Miami Beach, the authorities ordered an evacuation. My family was able to shelter in the hospital where my wife worked. But hospital administrators, understandably, wouldn't allow pets. We had a two-year-old dog who I was tasked with keeping alive. So I found a nearby parking garage—open on all sides but otherwise built like a fortress—and parked on the third floor, high above even the most apocalyptic flood level. I had a cooler of food in the trunk, a flask of bourbon in the glove compartment, a fully-charged laptop and plenty of work to do. I parked at the open edge facing north as the storm raged in from the west, and I watched as striated bands of clouds strayed across the sky and gradually thickened into a dense mass overhead. The wind surged, the rain sprayed the land in a torrent. The rain, in fact, was so strong, and came in such violent staccato bursts, it felt like I was being personally attacked. In the end, the center of the storm remained on Florida's west coast, and Miami suffered "only" a tropical storm with Category 1 and 2 gusts of wind. I wrote most of Chapter 6, "We're Going into the Storm," under the spell of that wind and water. The dog, lucky beast, slept the whole time.

But it wasn't until after the book came out that I wrote to the Air Force Reserve asking permission to ride a hurricane hunter. That was in February 2018, well in advance of the June to November hurricane season. Six months later, on Tuesday, October 9, a lieutenant colonel in the public affairs department called and informed me that if I could get to Biloxi by 6:15 the following morning, they would have room on a flight for me. With no margin for error, I bought a ticket to New Orleans (there are no direct flights to Biloxi from Miami) that arrived at midnight, rented a car and sped two hours east. After a few wrong turns in the dark, I finally managed to find Keesler Air Force Base at almost 3 A.M. Rather than rent a hotel room for a couple hours of sleep, I just curled up in the car. After Hurricane Irma, I was used to it.

The 53rd Weather Reconnaissance Squadron provides an invaluable government function and I wish I had spent more time on it in the book. The unit has its origins back in the 1940s, when pilots training in Texas during World War II dared each other to fly into an oncoming hurricane. They did it, and survived. Possibility became a reality with a tangible purpose. The planes could learn valuable information about a hurricane's size, power, and direction. Today, the squadron's main mission is to launch dropsondes, cardboard tubes containing data-gathering instruments, through a chute out the plane's belly into the storm's eye wall. The dropsondes collect information on moisture content, wind speed and direction, temperature, and atmospheric pressure inside the hurricane. This information is transmitted in real time to the National Hurricane Center, which uses the data for forecasts and advisories.

In other words, the hurricane hunters are the original source for much of the information that the NHC uses to warn people where storms are headed and how intense they are. As the aircraft commander on my flight, Lt. Col. Sean Cross, told me, "The data that we collect is very important. It's highly valued because, bottom line, we are saving lives."

And fast-growing Hurricane Michael was a life-threatening storm. Michael was powering up quickly, traveling over Gulf of Mexico waters that were 83 to 84 degrees Fahrenheit, three to four degrees warmer than average for that time of year in that part of the Gulf. If you've read this book, you know that warm water is hurricane fuel. The warmer the water, the faster storms develop. Michael went from a tropical storm blowing 35 miles per hour on October 8 to a Category 4 behemoth blowing 140 miles per hour seventy-two hours later.

The meteorologists following the storm were stunned.

"I'm in disbelief at how this intensified right at landfall, so late in the year, in this part of Gulf, so quickly," Ryan Truchelut, a meteorologist I'd interviewed for the book, told me.

Before heading to the plane, the aircraft commander, Lt. Col. Cross, briefed me (and five other civilian passengers) in the lobby of the air base. We were free to walk around the plane, but when a crew member motioned with their hands, we had to sit and buckle up immediately. The plane would be loud, so they handed out ear earplugs to protect our hearing. It might get bumpy, so they also handed out airsickness bags. This would be the 53rd's ninth and final mission into Michael. The storm was forecast to make landfall on the Florida Panhandle while we would be in the air. Once a storm makes landfall, the hurricane hunter's mission ends. The planes can't launch dropsondes over land; they might hit someone.

We clambered inside and took our seats on fabric-covered benches along the sides of the fuselage, then buckled in. It was going to be a five- to six-hour flight, so we had been told to bring our own food and water. The crew on this flight included Cross and his copilot, Maj. Dave Gentile; a backup pilot, Lt. Col. Byron Hudgins; two navigators, Christopher Harris and Kelly Soich; aerial reconnaissance weather officer Jeremy DeHart; and two loadmasters, Master Sgt. Chris Becvar and Staff Sgt. Jesse Jordan, who manned the dropsondes and other equipment in the rear of the plane. The planned itinerary was to fly at 10,000 feet into the storm and make four passes through the eye of the hurricane. At about 8:15 A.M., the plane roared down the runway and lifted off.

The ride up was, as promised, loud. But it was amazingly stable, given the winds we were flying into. The pilots explained that once they pick a spot to enter the storm, ideally where there's not a lot of lightning, they have to "crab" the plane, meaning fly at an angle to compensate for the force of the oncoming wind.

At altitude we were in the froth of the storm, gliding through clouds that enveloped us in their billows. As we penetrated deeper into the mass, the clouds condensed until there was nothing to see outside, just a wall of gray. I turned my attention inside, to Sgt. Jordan, who was staring at a radar screen showing green cloud cover over a black background. He was plotting the coordinates for the first dropsonde launch. Eventually he got up and slid the cardboard tube into the metal chute behind him. Minutes later a pop loud enough to be heard over the engine's roar announced the device's launch into Michael's eye wall. Because of this dropsonde, the National Hurricane Center was about to update its information on Michael's strength and speed and send it out to some very nervous people down below.

We made two circuits through the eye over the next couple of hours. But it wasn't until we made our third approach that the pilots invited me into the cockpit. I made my way unsteadily through the fuselage to a short ladder and climbed up into the cockpit. Inside, pilots Cross and Gentile sat at the controls. The navigator, Major Harris, sat at his own radar console. Backup pilot Hudgins and backup navigator Soich perched on jump seats.

Cross shouted above the plane's roar that we would be break-ing through the eye wall into the eye of the hurricane in about five minutes. Then he pointed at the radar, which showed the plane passing through the middle of the clouds and approaching an empty space on the screen. I looked out the window at a gray so solid and impenetrable it was blinding. There was no discern-ible orientation for up or down. The plane rumbled and vibrated, but as I stared outside we might as well have been standing still. I couldn't detect any forward motion. And then . . .

With a jarring suddenness, we broke through the cloud wall. The sun, a white-yellow orb, shone in a deep blue sky above us, even as an impressively puffy and undulating wall of clouds sur-rounded us. It was as if we were in a Renaissance painting of Heaven. The pilots and navigators were as stunned as I was, mur-muring "Wow," and "It doesn't get any better than this." As the plane banked to fly around the eye, we could see the white-capped and roiling waters of the Gulf beneath us. We were two miles up in the sky in the calm center of a Category 4 hurricane.

Hudgins, the backup pilot, stepped up to look out the win-dow. "I've been doing this twenty years, and I've probably seen that two or three times," he said. What impressed Hudgins was the barrel shape of the eye wall, a straight vertical cylinder, in-stead of the classic funnel or stadium shape. The absence of typ-ical gradation, he explained, was an indicator of the powerful convection throughout the storm's cloud deck.

At that moment, at that spot in the storm, Michael's winds were blowing 133 miles per hour. The pressure, measured in millibars—the lower the stronger—was an extremely low 922. But the storm wasn't fully developed. The winds would get stronger and the pressure would continue to drop. Clouds now stretched across 235 miles, with an eye of calm in the center 18 miles wide. Hurricane hunters like this one, maybe even this very one, had flown through Joaquin. Their crews were too deep in the clouds to see anything, but they nonetheless radioed out to

El Faro, over and over again, hoping to hear something back. They never did.

Having taken in the spectacle, I climbed back down the ladder and made my way to my seat. That's when things got bumpy. As I settled onto the bench, I reached into my luggage for some almonds before buckling in. Suddenly, there was a shudder throughout the plane, a light rumbling and shaking. In less than a second I was airborne. My ass lifted a good six inches off the seat and my legs flopped in the air like a kite tail in the wind. Around me, pieces of luggage shot up a foot off the floor. I instinctively clawed behind me to grab a handful of the mesh net that covered the inner wall of the fuselage. Then, just as quickly, I thumped back into my seat and it was over. I sat there stunned, as my organs fell back into place in my chest. Apparently, as we were exiting through the eye wall, we hit a patch of turbulence and the plane dropped as much as 2,000 feet. Gentile had the controls. He was one of the newest pilots in the 53rd; this was only his fourth hurricane mission.

"The section that we were going into was a really rough patch of turbulence. The autopilot kicked off, and you basically have to hand fly it," Gentile told me later. "And that's, I guess, what they pay us for."

Or, as Lt. Col. Cross put it, "Hurricane Michael had a hold of us and he was doing what he wanted to with the plane. For about thirty seconds there, it was pretty intense. It was a ride I'm not going to forget for the rest of my career."

In another hour or so, we made our final pass through the eye. Then, in a burst of excitement, the crew told us to look out the windows. As we gazed two miles below we could make out whitecaps, undulating ribbons of white foam where water met shore. We could see the waves hitting the sandy beach, and the green land of the Florida Panhandle behind it. This was the exact moment of landfall. It was a milestone for the hurricane hunters, whose missions routinely end before the storm reaches land.

"I don't know anybody else in the squadron that has seen that as long as I've been doing it," said Cross, who has been flying into hurricanes for eighteen years.

The storm was a glory to behold, a marvel, solidly constructed and stretching from heaven to earth. Yet we knew that all that power and beauty meant one thing: imminent destruction below. What appeared tiny and delicate to us was, in fact, a tremendous storm surge with waves reaching up to 20 feet crashing onto the shore, drowning whole beachside communities.

Michael slammed into land at 155 miles per hour, just 2 miles per hour shy of the Category 5 threshold, atomizing homes and buildings in its path. It was among the four most intense hurricanes to hit the continental United States since record-keeping began in 1851. The last dropsonde launched recorded a central pressure of 919 millibars, the third lowest reading for a hurricane that hit the continental U.S. Michael's stats may even be upgraded after the NHC reviews the data for its final report. As I write this, at least thirty-six people have died. Three more than perished on *El Faro*.

I made a conscious effort in the middle of the flight, when I couldn't see anything but the disorienting gray outside, to close my eyes and think about *El Faro*. To take just a moment to acknowledge all those who lost their lives when the ship went down.

Not long ago, I talked again to some of the helicopter crew who rescued the sailors from the *Minouche*. This was after the notorious 2017 hurricane season, during which three powerful hurricanes came ashore, devastating Houston, Puerto Rico, and the Florida Keys. The men told me how busy they had been. That was one of the busiest seasons for the hurricane hunters too. At one point the squadron flew three simultaneous missions, which is the maximum regulations currently allow. It's not lost on me that if this trend of powerful storms making landfall continues, and in all likelihood it will, the job of those working the front lines of extreme weather will become increasingly important—and risky. We'll be relying on the hurricane hunters for information that, as Cross put it, will save lives. We'll be relying on Coast Guard sailors, airmen, and swimmers to rescue us directly from the rain, wind, and waves. Through all of this, they'll be putting their own lives in harm's way.

For this, they have my gratitude.

—Tristram Korten, January 2019

ACKNOWLEDGMENTS

I want to thank all those who lost family on *El Faro* and talked to me about a difficult and emotionally wrenching episode in their lives: Tina Riehm, Carla Newkirk, Glen Jackson, Richard Griffin, Laurie Bobillot, and Claudia Shultz. I hope they feel I have treated the memory of their loved ones with respect.

I'm extremely grateful for the generous time and patience of numerous individuals in the U.S. Coast Guard, including AMT Joshua Andrews, AST Ben Cournia, Lt. Dave McCarthy, and Lt. Rick Post; Captain Jason Ryan; Captain Richard Lorenzen (retired); and Captain Jason Neubauer, among many others.

This book could not have been written without Captain Renelo Gelera of the *Minouche,* who generously shared with me his time and knowledge. He was instrumental in helping to school me in the finer points of navigation.

I would also like to thank AB Kurt Bruer, who very openly

talked about the survivor's guilt that is driving him to make sure an accident like this doesn't happen again, and who patiently answered many questions about his experiences on *El Faro*.

The National Transportation Safety Board collected and curated the information from the *El Faro* investigation in an open and transparent fashion, and Peter Knudson, in the office of media relations, responded in a timely manner to every question asked.

The National Hurricane Center in Miami accommodated all my questions regarding Hurricane Joaquin, always patiently and thoroughly. Todd Kimberlain, formerly of the NHC, was an invaluable resource in preparing this book and helping me understand some of the science surrounding hurricane formation.

Journalist Susan Eastman's assistance helping me navigate and understand Jacksonville's shipping community was extremely helpful.

I'd like to thank the entire team at Random House who helped make this book a reality: Gina Centrello, Kara Welsh, Jennifer Hershey, Kim Hovey, Emily Hartley, David G. Stevenson, Grant Neumann, Cindy Murray, Allison Schuster, and Dennis Ambrose.

I would also like to thank my agent, Nathaniel Jacks, who sought me out years ago, then stuck with me while encouraging me to look for a book subject that appealed to me.

Finally, this book improved tremendously under the eye of my editor, Brendan Vaughan, who worked on it tirelessly, diligently, and with passion. I owe him a debt of gratitude. A version of this story first appeared in *GQ* magazine, and likewise the editor who worked with me on it there, Geoffrey Gagnon, labored successfully to improve it. In addition, he gave it a platform that helped transform it into a book. Both of these editors have been encouraging and enthusiastic, which goes a long way toward sustaining an independent writer such as myself.

NOTES

The vast majority of the information in this book came from firsthand sources: interviews, testimony, official records, and observation. Any direct quotations were words said to me during an interview, captured on audio recordings, or spoken by witnesses during testimony. Any quotes or thoughts in italics represent the recollection of one party.

The sinking of *El Faro* was as thoroughly documented as possible by the National Transportation Safety Board and the U.S. Coast Guard's Marine Board of Investigation. Numerous witnesses testified, thousands of pages of documents were gathered, and two final reports were released. Most remarkable was the voyage data recorder, which captured five hundred pages' worth of heartbreaking conversations from the bridge of *El Faro*. I interviewed family members of all the sailors on the bridge except for two. Captain Michael Davidson's widow, Theresa, did not respond to a request through her lawyer to be interviewed. The widow of able seaman Frank Hamm also declined to be interviewed, as did the owners of *El Faro*, TOTE Inc., and its subsidiaries.

The sinking of the *Minouche* generated far fewer sources of official information, or even documented information. Except for some Coast Guard case reports and the ship's log, shared with me by Captain Renelo Gelera, I relied on interviews. I interviewed Captain Gelera, chief mate Henry Latigo, and able seaman Jules Cadet of the *Minouche;* the Coast Guard personnel involved in the rescue mission from Air Station Clearwater; Captain Jason Ryan of the Coast Guard cutter *Northland;* various personnel at the Coast Guard's District 7 Command Center; and Ranjit Gokhale, the captain of the Good Samaritan ship *Falcon Arrow.* Neither the owner of the *Minouche*, Milfort Sanon, nor the ship's agent, Caribbean Ship Services, responded to requests to be interviewed for this book. I visited Air Station Clearwater, Florida; Great Inagua, Bahamas; and the Coast Guard rescue swimmer school in Elizabeth City, North Carolina.

For information about Hurricane Joaquin, I interviewed personnel at the National Hurricane Center and read the NHC's archive of advisories and discussion posts for Hurricane Joaquin, as well as the center's Tropical Cyclone Report on the storm.

Supplemental information on hurricane science and history came from interviews with NHC scientists; Todd Kimberlain, who is no longer with the NHC; and academic scientists including Dr. Kerry Emanuel, among others. I also relied on Emanuel's *Divine Wind: The History and Science of Hurricanes* (2005); Erik Larsen's *Isaac's Storm: A Man, a Time, and the Deadliest Hurricane in History* (1999); *The American Practical Navigator,* originally by Nathaniel Bowditch and first published in 1802; and a 1927 article in the *American Mercury* by Willis Luther Moore, chief of the U.S. Weather Bureau in the 1890s.

Information on climate change's potential impact on hurricanes came from more than a dozen peer-reviewed studies published in scientific journals.

For background information on the merchant marine industry, I interviewed industry experts, studied trade journals, and relied on books such as Rose George's *Ninety Percent of Everything: Inside Shipping, the Invisible Industry That Puts Clothes on Your Back, Gas in Your Car, and Food on Your Plate* (2013), and Marc Levinson's *The Box: How the Shipping Container Made the World Smaller and the World Economy Bigger* (2006), about the shipping container revolution.

CHAPTER 1: CLEARWATER

The personal lives of the Coast Guard helicopter crew came from interviews with crew members and their wives. I visited Air Station Clearwater, toured the facilities, and flew on a Coast Guard C-130 from Clearwater to Great Inagua, Bahamas. For information about the C-130 Hercules, Coast Guard pilots briefed me and I reviewed the plane's technical specifications. I also read and enjoyed a *Popular Mechanics* article titled "Why the C-130 Is Such a Badass Plane," by Eric Tegler, published in April 2017. For information about Great Inagua, I consulted the Bahamas Chamber of Commerce, the Morton Salt company, and a historical study of labor unrest on the island titled "The 1937 Riot in Inagua, The Bahamas," by D. Gail Saunders, published in the *New West Indian Guide* in 1988.

CHAPTER 2: TROPICAL DEPRESSION ELEVEN

For the section aboard the *Minouche,* I relied on interviews with Captain Gelera, commercial ship information from Marinetraffic .com, Florida business incorporation records, and information from the Biscayne Bay Pilots association. For the section on the National Hurricane Center, I relied on interviews with James Franklin, the former branch chief of the NHC's Hurricane Specialist Unit; Robbie Berg, hurricane specialist; Daniel Brown, acting branch chief; and Todd Kimberlain, former NHC hurricane specialist. I used the NHC archive of advisories for moment-by-moment information about the tropical low.

CHAPTER 3: MORNING COLORS

The section on Air Station Clearwater relied on a series of interviews with Captain Richard Lorenzen, who is now retired, and Commander Scott Phy, with additional background information provided by Lieutenant Commander Thomas Huntley. Information for the section on the Miami River came from interviews with Captain Renelo Gelera, my own observations of cargo loading on these docks, photos of cargo loading in Port-de-Paix, Haiti, and the Biscayne Bay Pilots association. The historical section on the cargo industry came from Marc Levinson's book *The Box,* and Rose George's book *Ninety Percent of Everything*. The anecdote about Scottish cod being shipped to China came from an August 2009 article, "Scotland to China and Back Again," in Scotland's *Sunday Herald*. The history of the *Minouche* came from Coast Guard inspection and MISLE (rescue activity) reports and commercial ship tracking sites such as Marinetraffic .com and Fleetmon.com. For information about the world commercial shipping fleet, I referred to the 2016 UN Conference on Trade and Development report "Review of Maritime Transport." For the section on the practice of ships flying flags of convenience, I relied on general history accounts, news articles, and an interesting piece on the blog *Shipping Law Notes* by A. K. Febin, a scholar at the National University of Advanced Legal Studies in Cochin, India. The travails of the commercial shipping industry are well documented by news outlets including the *Los Angeles Times* and the BBC. Information on climate change came from "Atlantic Hurricane Trends Linked to Climate Change" (2006), in *Earth and Space Science News,* a journal of the American Geophysical Union; "Tropical Cyclones and Climate Change" (2010), in *Nature Geoscience;* and "Climate Forcing of Unprecedented Intense-Hurricane Activity in the Last 2000 Years" (2014), in

Earth's Future, also a publication of the American Geophysical Union. Information on waves came from studies in *J-STAGE,* a publication of the Japan Science and Technology Agency; the American Geophysical Union journals *Geophysical Research Letters, Earth's Future,* and *Earth and Space Science News; Coastal Engineering;* and the *Journal of Marine Science and Engineering.*

CHAPTER 4: PORT OF JACKSONVILLE

The history of Sun Shipbuilding and the building of *El Faro* came from the Sun Ship Historical Society, the Coast Guard's incident report, and TOTE Inc. The economics of shipbuilding came from an interview with shipping analyst Court Smith. *El Faro*'s in-port information came from the Coast Guard's Marine Board of Investigation report, and Port of Jacksonville docking records. The conversations quoted came from interviews with the river pilot Eric Bryson, and from testimony given by family members to the MBI, supplemented by interviews with family members. Information on the price-fixing conspiracy involving the Sea Star Line came from U.S. Department of Justice statements, documents filed in federal court, and a very good 2012 *Forbes* article by Walter Pavlo titled "Antitrust on the High Seas." Information on the captain fired by TOTE came from MBI testimony. Information on Captain Davidson came from the testimony of his wife and TOTE Inc. Second mate Danielle Randolph's concerns about *El Faro* came from the testimony of two friends, Korinn Mowrey and Norrie Thompson, and interviews with Randolph's mother, Laurie Bobillot. Information on helmsman Jack Jackson came from his brother, Glen Jackson. Information about Keith Griffin came from the testimony of his wife, Katie Griffin, to the MBI, and from interviews with his brother, Richard Griffin. Information about Richard Pusatere and Dylan Meklin came from testimony given by the relatives of both men to the MBI. Information on the low-pressure system came from the NHC archives for Hurricane Joaquin.

CHAPTER 5: THE OLD BAHAMA CHANNEL

Information about Tropical Depression Eleven and Todd Kimberlain came from interviews with Kimberlain and the NHC archives. Information about the *Minouche* and Captain Gelera came from the ship's log and interviews with Gelera. Information on the *Falcon Arrow* came from interviews with Captain Ranjit Gokhale, along with vessel information from the owner, Gearbulk, and Marinetraffic.com. Information on the Coast Guard cutter *Northland* came from the Coast Guard, the Coast Guard newsletter *Fair Winds,* and interviews with Captain Jason Ryan. Coast Guard history came from various publications, including editions of *Coast Guard Outlook* magazine, and an official history at forcecom.uscg.mil. The line about a hurricane powering the electrical grid comes from calculations done by Dr. Emanuel. Source material for hurricane history came from Kerry Emanuel's book *Divine Wind;* NOAA.org; the Powhatan Museum of Indigenous Art and Culture; and Erik Larson's *Isaac's Storm.*

CHAPTERS 6 AND 7: "WE'RE GOING INTO THE STORM" AND THE SAILOR'S DILEMMA

Information, including quotes from crew members, came from *El Faro*'s voyage data recorder, cross-referenced with information from the NHC's Joaquin archives and the Coast Guard's MBI report. Personal information about crew members Steven Shultz, Jeremie Riehm, and Jack Jackson came from interviews with Claudia Shultz, Tina Riehm, and Glen Jackson. Information about Frank Hamm came from interviews with former colleague Kurt Bruer and the article "*Horizon Producer* Rescues Stranded Fishermen" in the April 2011 issue of the *Seafarers Log,* a newspaper of the Seafarers International Union. Information on the *Novia Scotia* and *Titanic* accidents came from the book *Normal Accidents,* by Charles Derrow (1984).

CHAPTER 8: THE SECOND EYE

Information about District 7's Command Center came from interviews with operation unit controller Matthew Chancery and Captain Todd Coggeshall (now retired) as well as a tour of the facility. Quotes from the phone call between Chancery and Captain Lawrence of TOTE came from a transcript of the recording. Information about the *Emerald Express* came from Coast Guard records, confirmed by the company that owns the *Emerald Express*, G-G Marine, and a November 2015 article titled "Ship Beached after Hurricane," in *The Tribune*, a Bahamian newspaper.

CHAPTER 9: "FLY SAFE"

Information about Hurricane Ike came from the NHC report about the storm. Information about the Coast Guard's evacuation from the Bahamas and the rebuilding process came from interviews with Captain Richard Lorenzen (now retired) and Oldcastle Precast, the company the Coast Guard hired to do the rebuild. Information on preparations by the Jayhawk crew came from the crew members—Joshua Andrews, Ben Cournia, Dave McCarthy, and Rick Post—as well as Commander Scott Phy. Coast Guard aviation history came from various Coast Guard histories, including the aviation centennial edition of the *Coast Guard Outlook*. Helicopter flight information came from interviews with pilots and the FAA's regulations on helicopter controls.

CHAPTERS 10 AND 11: BATTLE RHYTHM and THROUGH SURF AND STORM AND HOWLING GALE

Information on the first and second phases of the *Minouche* rescue came from interviews with Coast Guard helicopter crew members Joshua Andrews, Ben Cournia, Dave McCarthy, and Rick Post; aviation maintenance technician Todd Taylor; and *Minouche* crew members Captain Renelo Gelera, chief officer

Henry Latigo, and able seaman Jules Cadet. History of the Coast Guard rescue swimmer program came from the official Coast Guard history at www.uscg.mil/history, as well as a tour of the school in North Carolina and interviews with Senior Chief Scott Rady and Master Chief John Hall. Information on the *Marine Electric* sinking came from the USCG marine casualty report on the incident. Weather information came from NHC advisories. Information about the Coast Guard cutter *Northland* came from interviews with Captain Jason Ryan, as well as from an informal log he wrote.

CHAPTER 12: CLIMBING THE CLOUD DECK

Information about the notification of families about *El Faro*'s disappearance came from interviews with the family members. Information on planning the C-130 sortie into Joaquin came from interviews with Coast Guard personnel involved, including Lieutenant Commander Jeff Hustace, Commander Scott Phy, and Captain Richard Lorenzen. Details on the search planning in District 7 came from interviews with and the testimony of Captain Todd Coggeshall, along with daily reports on the search produced by the Coast Guard. Details about the Jayhawk sortie into Joaquin came from interviews with Lieutenant Kevin Murphy.

CHAPTER 13: "TOGETHER, AS ONE CREW"

Information on the weather came from the NHC's Hurricane Joaquin archive. Information on the search came from Captain Coggeshall's testimony and from an interview with him. Information on the Jayhawk flight where the deceased sailor was found came from the pilot and rescue swimmer's testimony (their names were redacted) as well as some of the other Coast Guard sources listed here. Information on the Coast Guard cutter *Northland*'s search efforts came from interviews with Captain Jason Ryan as well as Captain Ryan's handwritten account of events. Information on the Coast Guard's meeting with families came from the Coast Guard and Carla Newkirk, Tina Riehm, and Glen Jackson.

CHAPTER 14: PORT-DE-PAIX

Captain Renelo Gelera shared information about the crew's sojourn on Great Inagua during several interviews. Information on the *Etoile des Ondes* came from the September 2010 accident report by the United Kingdom's Marine Accident Investigation Branch. Ana Gelera spoke with me about learning of her husband's ordeal. The chronology of the search for the wreck of *El Faro* and the voyage data recorder came primarily from NTSB statements, along with photos and data from the NTSB's online docket of information on the accident.

CHAPTER 15: "I'M NOT LEAVING YOU"

The transcript of the voyage data recorder provided the quotes and information regarding *El Faro*'s final hour. This information was cross-referenced with the Coast Guard's MBI report, for details on what was happening to the ship and when. Weather information came from the NHC's Hurricane Joaquin archive.

CHAPTER 16: EIGHT BELLS

Information regarding the lawsuits came from court documents. Information about the Coast Guard's post-accident inspection process and interactions with the American Bureau of Shipping inspectors came from the Coast Guard's MBI report.

CHAPTER 17: THE MASTER'S JUDGMENT

Information for this chapter came from the Coast Guard's MBI report; interviews with Captain Jason Neubauer of the Coast Guard and Thomas Roth-Roffy, formerly of the NTSB; and the testimony of TOTE executives Philip Morrell and Peter Keller. Climate change information came from studies published in *Nature Geoscience;* the American Geophysical Union journals

Earth's Future and *Earth and Space Science News; J-STAGE,* a publication of the Japan Science and Technology Agency; and the National Academies of Sciences, Engineering, and Medicine's report "Attribution of Extreme Weather Events in the Context of Climate Change." Hurricane information came from NHC reports on the hurricanes mentioned, a study in *Nature Geoscience,* and interviews with Kerry Emanuel. Theresa Davidson's statement came from the testimony of her lawyer. Information about Carla Newkirk's visit with her son to the memorial for her father came from interviews with Newkirk.

INDEX

PHOTO: © KEVIN ROSE

TRISTRAM KORTEN is a magazine, newspaper, and radio journalist. His work has appeared in a wide range of publications, including *GQ*, *The Atlantic*, and *The Miami Herald*. His reporting has aired on public radio programs nationally and in Florida. He is the former editor of the Florida Center for Investigative Reporting and was a 2013 University of Michigan Knight-Wallace Fellow. A long time ago, he graduated from Colby College. He lives in Miami with his wife, two daughters, and a mutt named Misha. This is his first book.

Twitter: @TristramKorten

ABOUT THE TYPE

This book was set in Galliard, a typeface designed in 1978 by Matthew Carter (b. 1937) for the Mergenthaler Linotype Company. Galliard is based on the sixteenth-century typefaces of Robert Granjon (1513–89).